Proceedings in Life Sciences

Algal Development

Molecular and Cellular Aspects

Edited by
W. Wiessner, D. G. Robinson
and R. C. Starr

With 87 Figures

Springer-Verlag
Berlin Heidelberg New York
London Paris Tokyo

Professor Dr. Wolfgang Wiessner
Professor David G. Robinson, Ph.D.
Pflanzenphysiologisches Institut der Universität
Untere Karspüle 2
D-3400 Göttingen, FRG

Professor Richard C. Starr, Ph.D.
The University of Texas at Austin
Austin, TX 78712, USA

ISBN-13:978-3-642-72606-4 e-ISBN-13:978-3-642-72604-0
DOI: 10.1007/978-3-642-72604-0

Library of Congress Cataloging in Publication Data. Algal development. (Proceedings in life sciences). Proceedings of the Third Symposium on Experimental Phycology, held Sept. 1986 in Göttingen under the auspices of the Akademie der Wissenschaften in Göttingen. Includes bibliographies and index. 1. Algae – Development – Congresses. 2. Algae – Cytology – Congresses. 3. Plant molecular biology – Congresses. I. Wiessner, Wolfgang, 1933- . II. Robinson, David G. III. Starr, R.C. (Richard C.), 1924- . IV. Symposium on Experimental Phycology (3rd : 1986 : Göttingen, Germany). V. Akademie der Wissenschaften in Göttingen. VI. Series. QK564.3.A57 1987 589.3'3 87-9767.

Softcover reprint of the hardcover 1st edition 1987

2131/3130-543210

P R E F A C E

In recent years our understanding of the molecular basis of cellular development has advanced significantly. At the same time algae have become more and more favoured for a whole range of studies in related areas of biology. This prompted Prof. A. Pirson, formerly director of the Göttingen Plant Physiology Institute and a well-known expert on algal physiology, to suggest that we should devote our third Symposium on Experimental Phycology to molecular aspects of algal development. The meeting took place in September 1986 and was attended by a number of internationally renowned scientists. This book represents the proceedings of this meeting.

Once again the meeting was organized under the auspices of the Akademie der Wissenschaften zu Göttingen with the financial assistance of the Stiftung Volkswagenwerk. The generous support of both of these institutions is greatly appreciated. We would also like to take this opportunity to express our thanks to all those who were involved in the day-to-day operation of the meeting. Last, but not least, the editorial staff of Springer-Verlag deserve a special word of recognition for enabling this book to appear so soon after the meeting had taken place.

Göttingen, April 1987

W. Wiessner
D.G. Robinson
R.C. Starr

CONTENTS

C O N T R I B U T O R S

You will find the addresses at the beginning of the respective contribution

INTERACTION OF CIRCADIAN OSCILLATORS AND THE CELL DEVELOPMENTAL CYCLE IN *EUGLENA*

Leland N. Edmunds, Jr.
Department of Anatomical Sciences
School of Medicine
Health Sciences Center
Stony Brook, New York 11794, U.S.A.

Mechanisms proposed to control algal and microbial cell division cycles (CDC) often include the notion of timers and clocks (Edmunds, 1984a). Various types of *autonomous* biochemical and macromolecular oscillators have been proposed to underlie mitosis and other "landmarks" comprising the CDC (see Edmunds, 1984b, 1986; Edmunds & Laval-Martin, 1984). These range from those of the relaxation type, in which a single continuous variable accumulates or declines, triggering some event in the CDC when it reaches a critical threshold and then resetting to a baseline so that the cycle starts again, to those exhibiting limit cycle dynamics (Edmunds & Adams, 1981). In the latter, a central clock, characterized by a self-sustained oscillation of at least two continuously varying biochemical species, would maintain stable periodic behavior and coordinate the timing of events comprising the CDC despite transient perturbations in phase; it would continue to function even if mitosis and division were blocked. Circadian oscillators (CO's) may underlie the persisting circadian rhythms (CR's) of cell division or of "hatching," having longer periods of 20–28 h, frequently observed in populations of algae, fungi and protozoa (Edmunds, 1984b; Edmunds & Adams, 1981; Edmunds & Laval-Martin, 1984). Our goal is to determine the biochemical nature of the oscillator(s) presumed to couple to and interact with the CDC and to generate these overt rhythmicities.

Circadian Control of the Cell Division Cycle in *Euglena*

The circadian "gating" of CDC events perhaps has been most intensively investigated in the algal flagellate *Euglena gracilis* Klebs (strain Z), a highly rhythmic, circadian system which has been well characterized physiologically and biochemically. The *formal* properties of circadian clocks—particularly entrainability, persistence, phase-shiftability, and temperature compensa-

Algal Development (Molecular and Cellular Aspects)
Ed. W. Wiessner, D.G. Robinson, and R.C. Starr

tion—have been found to characterize the CR of cell division in this organism. Inasmuch as this periodicity serves as the overt reference rhythm for our biochemical analysis of the underlying oscillator, we now review the major lines of evidence (Edmunds & Laval-Martin, 1984) that implicate an autonomous clock in the control of the CDC in this unicell.

Entrainability. Photoautotrophic cultures of wild type *Euglena* (Z strain) can be synchronized (entrained) by 24-h light-dark cycles (LD) so that cell division is confined entirely to the dark intervals, in sharp contrast to the exponential growth (g =12–14 h) obtained in continuous illumination (LL). If one reduces the duration of the light interval (e.g., LD: 8,16) within a 24-hour framework, the amplitude of the division rhythm is reduced, and the average g of the culture is lengthened (implying that not every cell divides during each division burst). Similarly, "skeleton" photoperiods comprising the framework of a full-photoperiod cycle (e.g., LD: 3,6,3,12), non-24-h LD cycles (but only within certain limits), and diurnal temperature cycles will also entrain the rhythm. Finally, photosynthetic mutants of *Euglena* can be synchronized with diurnal LD cycles, providing that the average g is longer than 24 h, thereby effectively circumventing the problem of the dual use of imposed LD cycles: as an energy source, or "substrate," for growth, on the one hand, and as a timing cue (*Zeitgeber*) for the clock, on the other.

Persistence. A basic test for CR's (*sensu stricto*) is to determine whether they continue to freerun for a number of cycles following transfer of the organism to conditions held constant with respect to the major environmental *Zeitgeber* (light, temperature); characteristically, the period (τ) under such conditions only approximates 24 h, as might be expected of an imperfect biological clock. Indeed, rhythmic cell division has been found to persist for many days in the Z strain of *Euglena* autotrophically grown in dim LL and in photosynthetic mutants organotrophically cultured in LL or DD (constant darkness).

We have also observed "free-running" circadian division rhythms in a variety of higher-frequency LD cycles and even "random" illumination regimes. Such short-period cycles (T<<24 h), provide no information to the cells with regard to 24-h time, although they may modulate τ to some extent. Those high-frequency regimens having symmetric photo- and scotophases (e.g., LD: 1,1 or LD: 3,3) and properly constructed random regimes have proved particularly useful in that they afford an amount of light for growth during a 24-h time span identical to that received in a full-photoperiod LD: 12,12 entraining cycle yet elicit free-running CR's not only of cell division, but also of motility and of photosynthetic capacity and chlorophyll content.

Phase Shiftability. Another property of CR's is that their phase can be reset (or shifted)—as the result of the lengthening or shortening of one or more oscillations—by single light (or dark) or temperature signals. Both the sign and magnitude of phase shift ($\Delta\phi$) engendered by the signal are predictably dependent on the circadian time (CT) at which the perturbations are applied. Similarly, the phenomenon of phase perturbation is inherent in most, if not all models for CDC oscillators: The response of cells to external influences is often strongly dependent on the time of the CDC at which the agent is imposed.

We have derived a detailed, Type 0 ("strong") phase-response curve (PRC) for a *circadian* mitotic clock by utilizing photoautotrophic cultures of *Euglena* free-running in LD: 3,3. At different times throughout the 30.2-h CDC, 3-h light perturbations were imposed systematically during one of the intervals when dark would have fallen. Both $+\Delta\phi$ and $-\Delta\phi$ were found—apparently constituting true developmental advances and delays—with maximum values of about 11–12 h being obtained at CT 22–23 (the "breakpoint"); little, if any, $\Delta\phi$ occurred if the light signal were given between CT 6 and 12. The CT at which $-\Delta\phi$ or maximum $+\Delta\phi$'s were achieved corresponded, respectively, to the approximate position of the CDC during which division occurred (commencing at about CT 12) and to the first few hours of the G_1 phase when the free-running rhythm of photosynthetic capacity displayed the lowest values. Although light is needed as a "substrate" for photosynthesis and the progression of the CDC in photoautotrophically cultured *Euglena*, it serves a quite distinct and separable function in phase-shifting and entraining the circadian oscillator underlying the rhythm of cell division, whose cyclic sensitivity to light is reflected in the PRC.

Singularity Point. Theoretical studies have predicted (and have been experimentally confirmed in other systems) that a CO having limit cycle dynamics might be rendered arrhythmic—characterized by a phaseless, motionless state—by a critical pulse of a certain strength and duration given at a specific time (termed the "singularity point," S^*) in the circadian cycle. As stimulus strength is increased, the transition from Type 1 (weak pulse) to Type 0 (strong pulse) resetting is necessarily discontinuous at one special phase point corresponding to this unique S^*.

We have recently discovered that a 40- to 400-lux pulse given at CT 0.4 (the approximate location of the break-point, at about CT 23) during the free-running division rhythm in Euglena induced arrhythmicity, with the population's reverting to asynchronous, exponential growth. The intensity (I^*) of this annihilating pulse and the CT at which it was imposed were found to be quite specific: A 300-lux stimulus given at CT 21.5 merely generated a $-\Delta\phi$ of the same magnitude found for 7500-lux signals. Different degrees of asynchrony were observed as one approached the boundaries (CT,lux). Thus, the existence of this "critical pulse" and its corresponding S^* not only further supports the hypothesis that a CO regulates the CDC in Euglena but also suggests (though does not demand) that the pacemaker has limit cycle dynamics.

Temperature Compensation. A final, rather remarkable property of CR's is that their period—but not their amplitude—is only slightly affected by the ambient temperature over the physiological range. This is just what one would anticipate in a functional biological clock measuring astronomical time. In contrast, the duration of the CDC (i.e., g) is commonly thought to be highly temperature-dependent, and, indeed, this is true for Euglena also. The period, however, of the CO hypothesized to underlie rhythmic division in a cell population is conserved. We have observed that τ of the free-running rhythm of division in the P_4ZUL mutant in DD is temperature-compensated over a range of 7°C (14–21°C). Likewise, we recently have found in the Z strain cultured in LD: 3,3 a Q_{10} of 1.05, indicating that τ is virtually unaffected by changes in temperature over a 10°C range (22–32°C).

Other Considerations. This short review has demonstrated the key role of a CO in the control of the CDC in Euglena. Mitosis would not be an essential part of the oscillator but would lie downstream from it: Blockage of cell division should not stop the system from oscillating (Edmunds & Adams, 1981; Edmunds &

Laval-Martin, 1984). Thus, division would be a "hand" of the underlying clock. This hypothesis has been confirmed by two findings: (i) If the division rhythm (free-running in LD: 3,3) was stopped due to low initial levels of vitamin B_{12}, and if this inhibition subsequently was released by re-addition of B_{12}, rhythmic division started up again in phase with an unperturbed control; and (ii) If a pulse of lactate was given to a free-running culture, temporarily accelerating the CDC and overriding the CO, the phase of the rhythm when it was finally restored after the substrate had been depleted was that of the control. These results are consistent with those found for the in-phase restoration of rhythmicity in the P_4ZUL mutant free-running in LL by addition of sulfur-containing compounds to the medium.

How are the observed shortenings and lengthenings of individual CDC's (implied by the PRC for light signals) generated by a master oscillator at the biochemical and molecular level? The evidence formally demands that a clock of some sort predictably insert time segments into (time "dilation"), or delete them from (time "contraction"), the CDC. The CDC might be "programmed" by the coupling together in various combinations of a collection of timing loops of different lengths to form a flexible timer, or "cytochron" (Edmunds & Adams, 1981). Temporal loci, or control points, would exist along the cytochron track at which decisions would be made with respect to the addition or deletion of time loops by interaction with a CO. Thus, the cytochron and the circadian clock are posited to be functionally independent (although not necessarily entirely separate as to mechanism).

Biochemical Analysis of the Circadian Oscillator

The difficulty in distinguishing between the "hands" of circadian clocks (mechanism-irrelevant events) and the "gears" themselves (clock-relevant processes) has impeded efforts to elucidate their mechanism(s). Phase-shift experiments have been used in attempts to ascertain whether or not a given process is an integral part of a circadian clock on the rationale that a transitory perturbation of either the state variables or the parameters that may be used to characterize an oscillation can cause a permanent $\Delta\phi$ of an overt rhythm. Unfortunately, the converse is not necessarily true; an observed $\Delta\phi$ may have occurred as a result of the effect of the drug or other perturbing agent on some site only secondarily affected by the drug rather than on its postulated primary target (Goto et al., 1985).

Rationale. We have designated (Goto et al., 1985) any element as a "gear" (G) if it can be expressed as a state variable or parameter; if not, it is a non-gear (∿G). Together, the set of gears would constitute a closed control loop, or oscillator. Because an unperturbed ∿G oscillating in its physiological range would not be expected to regulate the operation of the G's themselves, its artificial perturbation within this range should not perturb circadian timekeeping (no steady-state $\Delta\phi$ in an overt rhythm would be observed). Consequently, if an experimental alteration in the level of a target within its normal range perturbs the clock and generates steady-state $\Delta\phi$'s, then that target is most likely a G (Criterion A). The activation and resulting increase in the level of a ∿G, however, might perturb timekeeping, whereas its inhibition would not (or vice versa). To more stringently differentiate between G and ∿G, therefore, a further requirement for a target to be classified as a G is that both the direct activation as well as direct inhibition of the target must perturb the clock (Criterion B). In principal, these criteria are applicable to any set of processed thought to constitute an autonomously oscillatory clock.

Oscillating Elements in Euglena. The in vivo levels of NAD^+, $NADP^+$ and NADPH were measured in synchronously dividing and in very slowly dividing cultures of Euglena grown photoautotrophically at 25°C in LD: 3,3 and were found to oscillate with a τ of 27 h and with an amplitude not directly related to the CDC (Goto et al., 1985). There was also a circadian rhythm in the activity of NAD^+ kinase (peak at CT 0) in extracts of Euglena with a phase relationship such that it could induce the rhythm in the in vivo level of NAD^+ (see Goto, 1984). No circadian oscillation of the ratios of either NADH/(NAD^+ + NADH) or NADPH/($NADP^+$ + NADPH) occurred. Finally, a 3-h light pulse applied at CT 18 to the free-running rhythm of cell division not only generated the expected $\Delta\phi$ of the division rhythm but also shifted the phase of the oscillation in total NAD^+ content by approximately the same amount (Goto et al., 1985). These results suggest that the circadian oscillation in the in vivo level of NAD^+ could be ascribable, as for Lemna (Goto, 1984), to that of the conversion between NAD^+ and $NADP^+$, but not to that of reduction—oxidation between NAD^+ and NADH.

To determine whether NAD^+ constitutes a clock G according to the criteria given in the rationale, small (25 ml) seed cultures displaying a free-running circadian rhythm of cell division were pulsed (for 2 h) at various CT's with NAD^+ (0.5 mM) or $NADP^+$ (0.2 mM)—expected to directly elevate their own in vivo levels and then to increase or decrease the rate of the reactions catalyzed by NAD^+ kinase and $NADP^+$ phosphatase— or p-nitrophenylphosphate (10 mM), a competitive inhibitor of $NADP^+$, and then resuspended in fresh medium (effectively terminating the pulse by dilution) for subsequent automated monitoring of the cell division rhythm. PRC's for the division rhythm could be derived with each of these agents (Goto et al., 1985). Furthermore, a 2.5-h pulse of NAD^+ given at CT 21.7, which caused a - ø of 4 h in the division rhythm, also phase-shifted the CR in the in vivo levels of of NAD^+, NADH, $NADP^+$, and NADPH. These results suggest, therefore, that NAD^+ (or $NADP^+$, or NADPH), NAD^+ kinase, and $NADP^+$ phosphatase represent G's of the underlying CO.

What, then, is the element that regulates these enzymes, already suggested to be G's in themselves? Inasmuch as Ca^{2+} -calmodulin activates NAD^+ kinase in many plants, including green algae, and in the sea urchin, and because the circadian rhythms in the activities of NAD^+ kinase and $NADP^+$ phosphatase appear to be generated by a rhythm in the in vivo level of this complex in *Lemna* (Goto, 1984), it would seem to be a likely candidate for this G in *Euglena* also. We attempted (Goto et al., 1985) to *directly* cause a transitory decrease in [Ca^{2+}] by means of 2- to 3-h pulses of chlortetracycline (200 μM), a membrane-permeable chelator of Ca^{2+}, or to transitorily inhibit Ca^{2+} -calmodulin by similar short pulses of W7 (20 μM) and chlorpromazine (50 μM), both calmodulin inhibitors. All three drugs yielded pronounced Δø's of the cell division rhythm; the PRC's obtained suggest that both cytosolic Ca^{2+} and calmodulin constitute G's.

If so, there should be another G directly regulating the level of Ca^{2+}. The main regulatory sites for many non-circadian systems are known to be the plasmalemma, the endoplasmic reticulum, and the mitochondria. To test the possibility that the mitochondrial Ca^{2+} -transport system might be a G, cultures were pulsed (2—3 h) with nitrogen (600 ml min^{-1}), dinitrophenol (100 μM), and sodium acetate (10 mM), all of which affect electron transport, ATP hydrolysis and mitochondrial Ca^{2+} efflux. Each of these agents phase-shifted the division rhythm (Goto et al., 1985). The mitochondrial Ca^{2+} -transport system, therefore, would appear to be a G of the oscillator in *Euglena*.

A Model for the Circadian Oscillator. Our findings suggest that NAD^+, the mitochondrial Ca^{2+} -transport system, Ca^{2+}, calmodulin, NAD^+ kinase and $NADP^+$ phosphatase represent clock "gears" that might constitute a self-sustained, circadian oscillating loop in *Euglena* and other eukaryotic microorganisms, as well as in higher plants and animals. We have proposed a model (Goto et al., 1985)—not for "the" clock—but for one oscillator in what is most probably a cellular "clock-shop." This regulatory scheme includes the following three steps: (1) NAD^+ (or a stimulated photoreceptor) would enhance the rate of net Ca^{2+} efflux from the mitochondria or net Ca^{2+} influx across the plasmalemma into the cytoplasm, resulting in a maximal concentration of cytosolic Ca^{2+} 6 h (90°) later; (2) Ca^{2+} would immediately form an activated Ca^{2+} -calmodulin complex in the cytoplasm (maximal level at 90°); (3) This active form of Ca^{2+} -calmodulin would decrease the *rate* of net production of NAD^+ by both activating NAD^+ kinase and inhibiting $NADP^+$ phosphatase in the cytoplasm so that the rate would become maximal 12 h later (at 270°) when Ca^{2+} -calmodulin reaches its minimum level. After 6 h more when the in vivo *level* of NAD^+ then becomes lowest (at 0°), the regulatory sequence would be closed, and the cycle would repeat.

Acknowledgement

Supported in part by National Science Foundation grants (PCM-8204368 and DCB-8501896) to L. Edmunds.

References

Edmunds,L.N.,Jr., ed. (1984a) Cell Cycle Clocks. Marcel Dekker, New York.
Edmunds,L.N.,Jr. (1984b) In: The Microbial Cell Cycle. eds. P. Nurse, E.Streiblová, CRC Press, Boca Raton,Florida. pp.209-230.
Edmunds,L.N.,Jr. (1986) In: Endocytobiology III. eds. J.L.Lee, J. F.Fredrick, Ann.N.Y.Acad.Sci. (in press).
Edmunds,L.N.,Jr.,K.J.Adams (1981) Science 211, 1002-1013.
Edmunds,L.N.,Jr.,D.L.Laval-Martin (1984) In: Cell Cycle Clocks. ed. L.N.Edmunds,Jr., Marcel Dekker, New York. pp.295-324.
Goto,K. (1984) Z.Naturforsch. 39c, 73-84.
Goto,K.,D.L.Laval-Martin,L.N.Edmunds,Jr. (1985) Science 228, 1284-1288.

CONTROL POINTS IN THE *CHLAMYDOMONAS* CELL CYCLE

P.C.L. John
Department of Developmental Biology
Research School of Biological Sciences
Australian National University
GPO Box 475
CANBERRA CITY ACT 2601

Chlamydomonas has the crucial advantage, for genetic analysis of the cell cycle, of being haploid. The experimental convenience of being readily synchronised in division by alternating periods of light and dark, raises the question of how such synchrony arises and whether it indicates that the cell cycle in *Chlamydomonas* is fundamentally different from that of most eukaryotes. We have earlier argued that it is not fundamentally different (John 1984) and have proposed that at least a functional similarity exists between the "start" division control of yeasts (Nurse 1985) and the attainment of commitment to divide in *Chlamydomonas*. In this brief acount we include further evidence in support of these propositions.

Timer Control

It has sometimes been assumed (Edmunds *et al.* 1981) that because division can be synchronised by light - dark alternation, therefore an endogenous circadian oscillator must be controlling division in *Chlamydomonas*, as it sometimes can do in *Euglena* (Edmunds, this volume). Alternatively Spudich and Sagar (1980) rejected timer control, and proposed that synchrony arises because there are metabolic events essential to different phases of the cell cycle that are directly dependent upon light or dark.

Our experiments, prompted by these stimulating hypotheses, lead us to support Spudich and Sagar in finding no detectable influence of an endogenous oscillator running through successive cycles. However, the hypothesis of specifically light or dark dependent processes is not supported by evidence that absence of CO_2 in light is equivalent to darkness in its synchronising effect. There is also clear evidence of timer operation within each cell cycle; though not running-on in endogenous oscill-

Algal Development (Molecular and Cellular Aspects)
Ed. W. Wiessner, D.G. Robinson, and R.C. Starr

ations and so not linking the timing of a current cycle to that of earlier ones.

Evidence against the operation of an endogenous circadian oscillator is of five sorts. First, daughter cells held in darkness show no oscillation in capacity to enter a new cell cycle (Spudich *et al.*1980; McAteer *et al.* 1985). Second, cells in continuous illumination can immediately initiate a new cell cycle, of normal duration timed from the formation of daughter cells, even if the previous cell cycle was very different from 24h duration. Differences of cycle duration result from incomplete stabilisation of the cycle against changes in growth rate caused by available light energy, and cycle length conforms to equation (2) of Donnan *et al.* (1983, 1985).

M.G.T. = 11 + 0.75 mass doubling time in hours ....(2)

therefore cycle lengths can differ widely from 24h. Third, normal cycle timing is also observed if the beginning of the cell cycle is delayed by slowing of the rate of progress through division processes in the previous cycle by low temperature (McAteer *et al.* 1985). Fourth, cells growing at very slow rates (mass doubling times longer than 44h) have cycles far longer than 24h and of whatever length is necessary to obtain the critical minimum size necessary to start division, as shown in Fig.1. There is no gating of division to 24h intervals, as can occur with photosynthetic protozoa (Hastings *et al.* 1964). Fifth, cells can be synchronised by a period of CO_2 deprivation without darkness and so without the light-dark transition presumed to entrain an oscillator.

Although this is strong evidence against a significant effect of an endogenous oscillator form of timer, there is clear evidence for the operation of very effective timers within the cell cycle. The duration of the period from the beginning of growth in autonomous daughter cells up to the attainment of first commitment to divide, is temperature-compensated in the range between 20° and 30°, in spite of the halving or doubling in rate of metabolic processes and growth that results from such temperature change. The period from first commitment to the completion of division also recovers its duration after a lag of one cell cycle following temperature change in the same range. Thus the whole cell cycle shows the key property of biological timers; that of being temperature-compensated. However, two different timers are involved, since the one controlling the precommitment period is dependent upon growth and is influenced by the rate of growth; its duration in hours being 5 + 0.75 the mass doubling time (eq. (1) of Donnan *et al.* 1983).

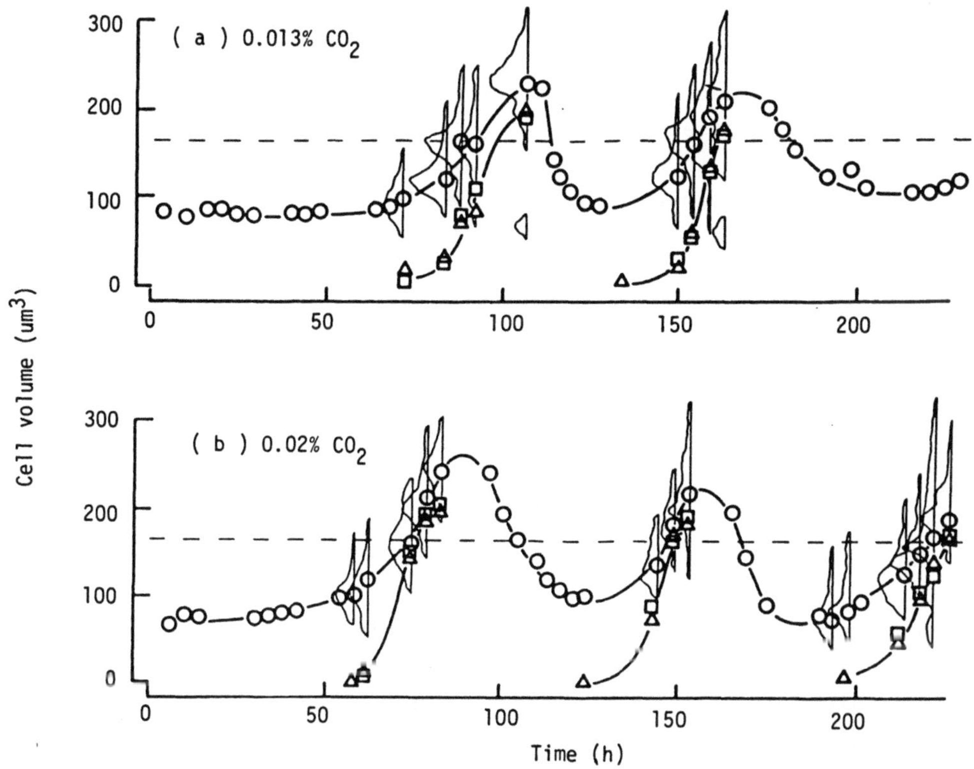

Fig. 1 The effect of (a) 0.013% CO_2 in air and (b) 0.02% CO_2 in air, on growth in volume and division, of synchronous cells transferred from time 0 to continuous illumination at 220 umol m^{-2} s^{-1} PAR. The cells had been synchronised by L/D during aeration with 0.5% CO_2 and were transferred to the experimental conditions at the end of a dark period. To measure commitment to divide samples of culture were transferred to continued aeration in darkness and allowed to complete any committed divisions. Evaporation from these commitment samples was checked by weight and a small correction applied if necessary. For measurement of cell volume, samples from the main cultures were fixed with formaldehyde and the cells were photographed and printed at x 710. Ellipsoid volumes were automatically calculated and collated, by a Videoplan analyser, which yielded size frequency profiles. These are shown as histograms with fine lines, in which the vertical abscissae share the same scale as the ordinates of the main graph, and the ordinates (extending leftwards but not shown) are: number of individuals in size class. These frequency profiles are shown for all samples in which members of the population were attaining commitment to divide, but at other times only the mean volumes derived from them are shown.

Plotted are; (O) mean cell volume; (Δ) proportion of the population committed to division; (□) proportion of cells exceeding a volume of 165 um^3. This critical volume is indicated by the broken horizontal line.

The critical volume was determined as a mean, derived from measurements at four different growth rates in populations progressing through two cell cycles. To derive the mean, the percentage commitment to division at each time was observed and in the cell size frequency profile a threshold volume was calculated; the proportion of cells above which matched the proportion committed to division. There was no pattern of change in this threshold as populations progressed through commitment. Mean critical threshold volumes, calculated in two cell cycles at four growth rates were; at 0.01% CO_2, 168 and 168 um^3 ; at 0.013% CO_2, 164 and 162 um^3 ; at 0.017% CO_2 156 and 161 um^3 and at 0.02% CO_2, 173 and 166 um^3, so indicating an overall mean of 165 um^3 as the critical minimum size for commitment under limiting CO_2.

Conversely, the time period after first commitment is little influenced by growth rate but is only adaptively compensated against temperature change. Since each timer can begin on the stimulus of completing the segment of the cell cycle controlled by the other timer (Fig. 2), but each is not detectably influenced by the timing of events within earlier cell cycles, they can be classified as hourglass timers; recognising their similarity with a sandflow timer that measures from its inversion.

It is this timer control of cycle duration that necessitates multiple fission, since it schedules a division to occur after more than a doubling of cell size, at mass doubling times shorter than 44h. Multiple fission is not determined at a single commitment event. Rather the interruption of growth at successive closely-spaced time intervals in synchronous cultures has revealed that cells pass through a succession of separate commitments, each to a single doubling of cell number, to the maximum that mother cell size allows (illustrated in Fig. 2).

It is noteworthy that the hourglass form of timer, in this cell cycle which allows multiple fission, permits division to be timed for the approach of night even at higher growth rates. This is not possible with endogenous oscillator control in cells that divide into two. *Euglena* can, in rich medium, double its mass in less than 24h (down to about 12h) but at these higher rates must uncouple its cell cycle from oscillator control to balance division rate with growth and so stabilise cell size. Divisions continue successfully, so the oscillator is clearly not an essential part of the cell cycle.

Start and commitment of divisions

Both "start" of division in yeasts (Hartwell *et al.* 1977; Nurse 1985) and commitment in *Chlamydomonas* occur just prior to S phase and commit a single doubling of DNA and cell number without further growth requirement. A point of difference in *Chlamydomonas* cells growing at rates requiring multiple fission is that the first commitment is regulated by a timer rather than cell size (Donnan *et al.* 1985) as illustrated in Fig. 2. We have recently detected that this difference is conditional upon higher growth rates. In very slowly-growing cells, with mass doubling times longer than 44h, the timers would schedule division to recur in less than a mass doubling time (eqn. (2) above). We have then observed that a requirement for a critical minimum size becomes rate-limiting for commitment and it extends the cell cycle, so that mean generation time is equal to mass doubling time.

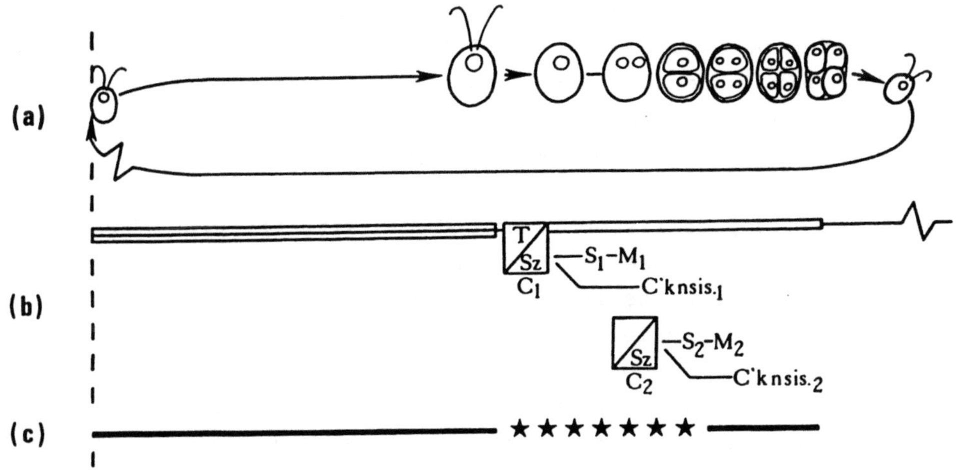

Fig. 2

(a) A summary of structural events in a cell cycle producing four daughter cells. The control point for each event is illustrated vertically below it. The cycle is timed from the initiation of growth in completed daughter cells (¦) and this initiation may be delayed until the end of a dark period (/\/).

(b) A major rate-limiting control point is the attainment of commitment to divide (C) for which concurrent growth is necessary, but after which division will occur with no requirement for further growth or the presence or absence of illumination. Additional commitments can occur, each to a single further round of DNA replication and doubling of daughter cell number. Each commitment initiates at least two sequences of events; one sequence concerning the nucleus in which mitosis (M) is dependent upon completion of DNA synthesis (S), another sequence in cytokinesis (C'knsis.) involving deployment of cytoskeleton and development of cleavage furrow. Progress within each sequence is independent of progress in others but all must be completed for successful division. Attainment of first commitment (C_1) requires expiry of a temperature-compensated timer (≡) and the attainment of a critical minimum size (T/Sz). This size requirement is cryptic at higher growth rates because of the extent of growth in the timer period. In larger cells, commitments can recur while the current mother cell size represents an apportionment per currently committed daughter cells above 27 pg protein; thus the minimum inherited mass is 13 pg and the maximum is 27 pg protein, but these limits are broadened by sloppy divisions.

(c) Division synchrony is induced by periodic growth because the beginning of growth in daughter cells is followed by a timed period in which no commitments to division can occur (——). This phase is followed by a relatively brief period in which commitments to a doubling of cell number can recur while mass per committed daughter is above a critical minimum (★★), resulting in a subsequent wave of division activity. There is then a reversion to the alternate state, in which the commitment timer must be initiated and run to completion before new divisions can be started.

We were able to obtain mass doubling times longer than 50h by the use of light intensities below 15µmol $m^{-2}s^{-1}$ PAR, or CO_2 concentrations at and below 0.02% in air. Such growth-limiting conditions must often occur naturally but we are not aware of previous reports of division studied under these conditions. We observed in cells limited by CO_2 that the proportion of cells in a synchronous population that had attained commitment to divide, consistently correlated with the proportion that had exceeded a critical size of 165µm^3 and that the critical size was not modulated by differences in growth rate between 80h and 160h mass doubling time, with the same limiting environmental factor. Fig. 1 shows the behaviour of daughter cells transferred from the same population to continuous illumination and, either, (a) 0.013% CO_2 in air, or (b) 0.02% CO_2 in air. The size frequency of cells in culture samples (fine lines) can be compared with the critical volume of 165µm^3 (dashed lines). The proportion of cells that exceeded the critical volume (□) correlated with that which had attained commitment to divide (▲). The correlation was confirmed at two other growth rates.

The commitment timer was not inactivated by slow growth rates. At a growth rate so slow that small daughter cells had to extend their cell cycle to attain the critical size, large daughter cells, which attained the critical size ahead of the scheduled time by virtue of a large initial size, were observed to delay commitment until expiry of the timer. Therefore we conclude that two conditions must be met before commitment can occur. These are: expiry of the timer, and attainment of a critical minimum size. Whichever is last attained becomes the rate-limiting requirement and the other requirement is cryptic. At the conventional high growth rates of laboratory cultures, the critical size has been exceeded by the time of commitment, so this requirement has not previously been detected.

At low growth rates, the *Chlamydonomas* cell cycle clearly resembles that of other eukaryotes since its duration is then equal to the mass doubling time, the start of division follows attainment of a critical size and only two daughters are formed.

Dependent sequences of events

It has become clear that, in budding yeast, the temporal ordering of the individual events that contribute to the process of division, results from their arrangement into dependent sequences; in which later events are only initiated after the completion of earlier events of that seq-

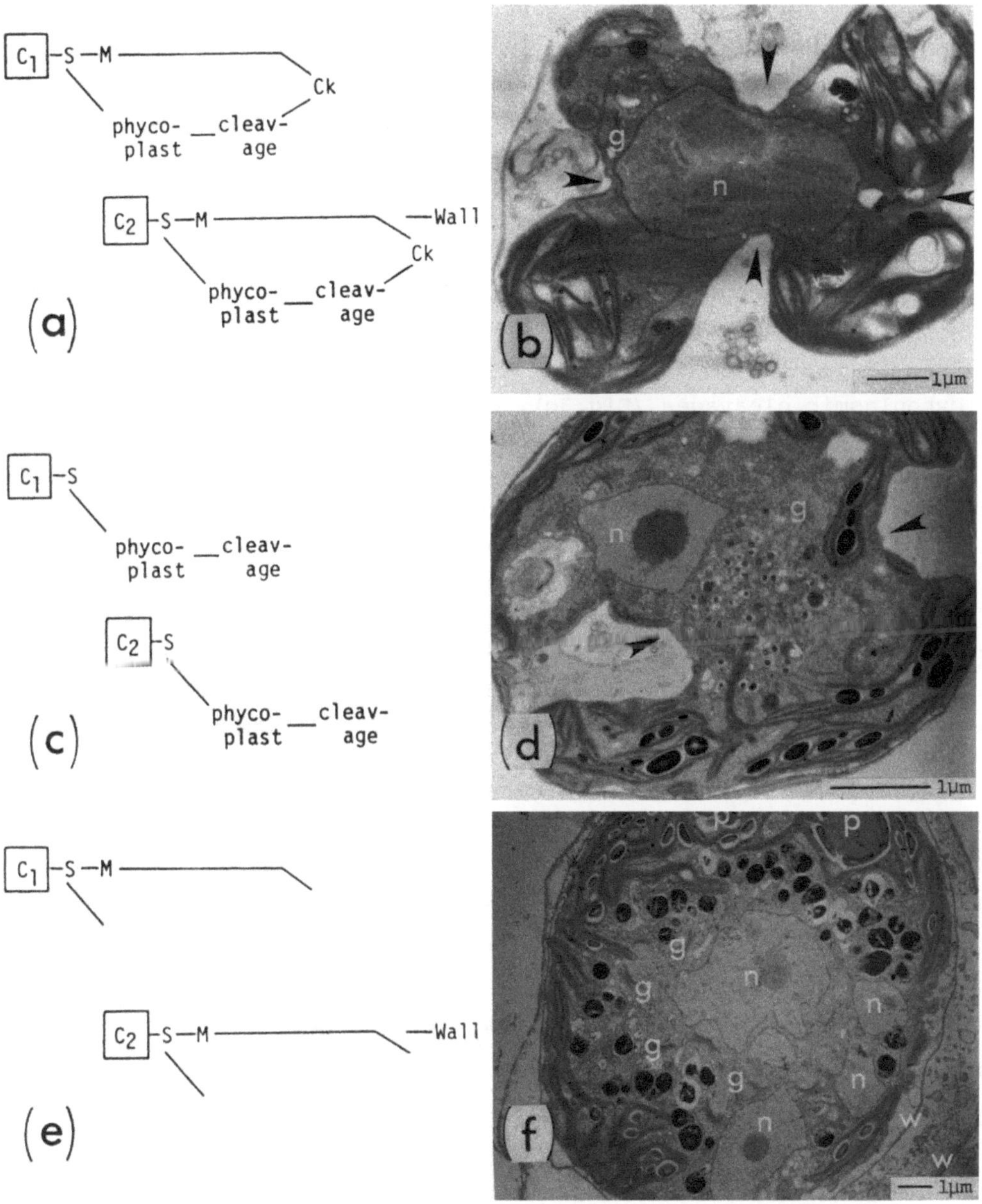

Fig. 3 Possible dependent sequences of events, and terminal arrest configurations resulting from their interruption.
Progress through a normal cell cycle is summarised in (a). The continuation of cleavage events in cells blocked in nuclear division by 2'-deoxyadenosine (b), or cdc mutation (d) are tentatively accounted for in (c). Conversely the continuation of nuclear division events in cells blocked in cleavage is predicted in (e) and a corresponding configuration, observed in oryzalin-inhibited cells, is shown in (f) note the multiple nuclei (n) and also the continued replication of pyrenoids (p), golgi (g) and wall layers (w) that are not dependent upon the cytoplasmic cleavage sequence of events, although their dependency relationships remain to be determined.

uence (Pringle *et al.* 1981). To test whether this principle extends to plant cells we have used metabolic inhibitors (Harper *et al.* 1986) and temperature conditional cell division cycle (*cdc*) mutants to interrupt particular division processes and test for independent progress in others. Both forms of arrest indicate that at least two separate sequences of events are initiated by each commitment. One sequence runs through DNA synthesis and mitosis, the other through deployment of the cytoskeleton and development of the cleavage furrow. Each sequence can proceed in the absence of progress in the other sequence. Cells blocked in nuclear division by inhibition of DNA synthesis with 2'-deoxyadenosine (Fig.3b), or in nuclear division by mutation (Fig.3d) can attempt one or more cytoplasmic cleavages (Fig.3c). Conversely, cells blocked in cytokinesis by the anti-tubulin agent oryzalin at 10 μM (Fig.3f) can execute the nuclear divisions to which they have become committed and so become multinucleate (Fig.3e). There are additional sequences, involving control of chloroplast division (Harper *et al.* 1986), and also pyrenoid division, which is clearly independent of cytokinesis (Fig.3d). There is a clear parallel with the taxonomically-distant budding yeast, in which separate bud emergence and nuclear division sequences are initiated at "start" (Pringle *et al.* 1981). Such separate dependent sequences may therefore be a fundamental feature of the organisation of division in eukaryotes.

LITERATURE REFERENCES

Donnan, L., E.P. Carvill, T.J. Gilliland, P.C.L. John (1985) New Phytol. 99,1-40.

Donnan, L., P.C.L. John (1985) Nature 304, 630-633.

Edmunds, L.N., K.J. Adams (1981) Science 211, 1002-1013.

Harper, J.D.I., P.C.L. John (1986) Protoplasma 131, 118-130.

Hartwell, L.H., N.W. Unger (1977) J. Cell Biol. 75, 422-435.

Hastings, J.W., B.M. Sweeney (1964) In: Synchrony in Cell Division and Growth. ed. E. Zeuthen, Interscience, New York pp.307-321.

McAteer, M., L. Donnan, P.C.L. John (1985) New Phytol. 99, 41-56.

Nurse, P. (1985) Trends In Genet. I, 51-55.

Pringle J.R., L.H. Hartwell (1981) In: The Molecular Biology of the Yeast *Saccharomyces cerevisiae*. ed. N. Strathern, E.W. Jones, J.R. Broach, Cold Spring Harbour Laboratory Press pp.97-142.

Spudich, J.L., R. Sagar (1980) J. Cell Biol. 85, 136-145

TWO POSSIBILITIES FOR TIME MEASUREMENT IN SYNCHRONOUS CHLORELLA - CIRCADIAN RHYTHM AND TIMING

Rudolf Tischner and Harald Lorenzen

Pflanzenphysiologisches Institut, University of Göttingen, Untere Karspüle 2
3400 Göttingen, F.R.G.

Endogenous rhythms with periods of about 24 h (called circadian rhythms) are common to all eucaryotic organisms (Edmunds, 1984). It is still unknown how such oscillations have arisen during the evolution. Pittendrigh (1966) suggested that in earlier times organisms which perform their DNA replication and cell division during the night may have an advantage because of the high UV radiation during the day. This might cause deletions of DNA by the formation of thymidine dimers. Another still open question is that concerning the number of clocks in an, even unicellular, organism. One suggestion is the presence of one central oscillator which controls the circadian output of different parameters, so called hands. A desynchronization of these parameters recorded as phase shifts indicates that the coupling with the oscillator is not tight and can be affected by changes in the environment.

The experiments which will be presented here were performed with the Chlorella strains 211-8b and 211-8k respectively. The general design of the experiments is given in Fig. 1.

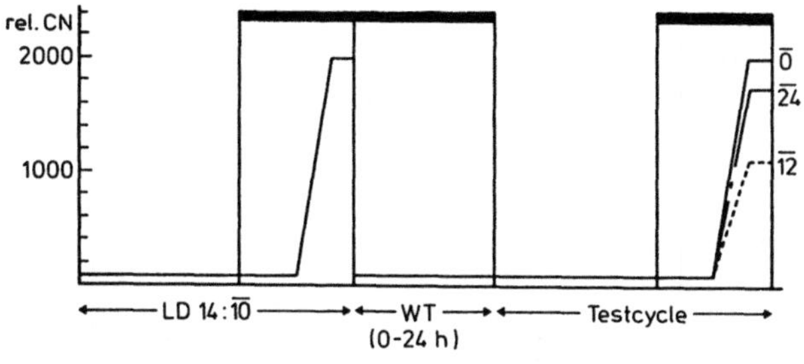

Fig. 1 The design of the experiments presented. The black bars indicate darkness of the light dark change. rel. CN: relative cellnumber, LD: light dark change, WT: waiting time.

Circadian oscillations in dryweight, carbohydrate (fig. 2), chlorophyll and protein content (data not presented) were observed. It should be recognized, that in order to measure these parameters the testcycle must not necessarily be completed (comp. Hesse, 1974). The oscillations were therefore monitored after only 14 h of illumination. However, if the autospore production is employed as parameter the complete 14:10 h light dark cycle is essential because the daughter cell release occurs in the dark (fig. 3) (Hesse, 1971; Lorenzen and Hesse, 1968).

Algal Development (Molecular and Cellular Aspects)
Ed. W. Wiessner, D.G. Robinson, and R.C. Starr

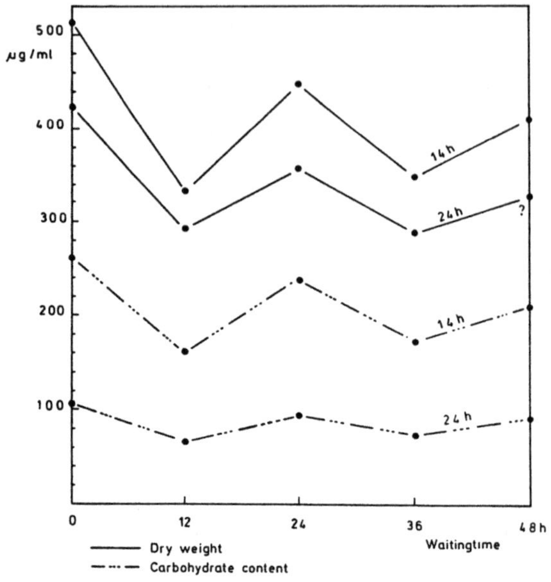

Fig. 2 A circadian oscillation in the capacity to produce dryweight and carbohydrates was obtained dependent on the waiting time (x-axis). The differences in production can be observed after 14 h illumination or after a complete 14:10 light dark change.

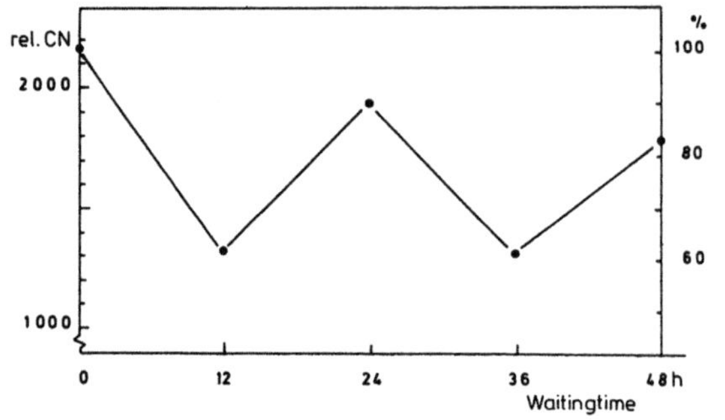

Fig. 3 A circadian oscillation in the capacity of daughter cell production dependent on waiting time (x-axis).

All data were obtained within a 48 h waiting period although we have followed these oscillations for more than 4 days with the well-known damp out effect. Such a fading of the oscillation was also observed if low light intensities of about 0.07 W/m^2 replace darkness during waiting time. Then the capacity to produce is fixed at the lowest level.

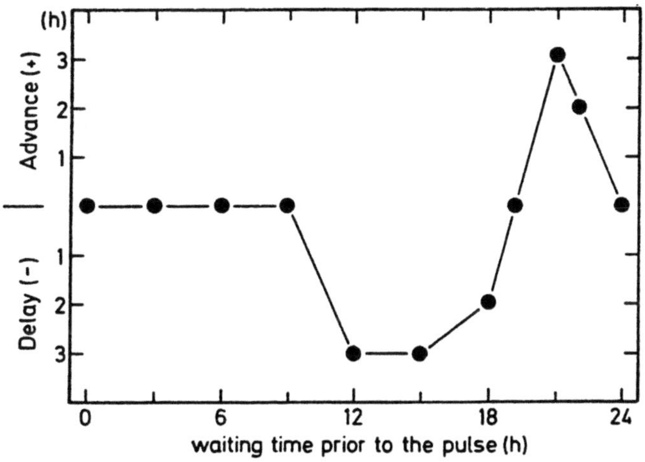

Fig. 4 Phase response curve of maximum productivity in Chlorella. The waiting time was interrupted by light pulses (15 W/m^2, 0.5 h) at different circadian time. The timing of maximum autospore production was the phase reference event.

From these data it becomes obvious that the capacity of the cells for the production of cellular compounds and finally daughter cells oscillates during darkness in a circadian way. The typical features of this oscillation (e.g. entrainment, persistence and temperature compensation) have been presented previously (Hesse, 1971). In addition Wu et al. (1986) have shown a phase response curve due to light pulses during a waiting time (fig. 4). The same result was obtained with pulses of white, red or blue light.

Therefore, it becomes clear that in Chlorella the capacity for biomass production is obviously regulated by coupling to a circadian oscillator. Surprisingly these oscillation can be measured even at the end of an additional generation time. This indicates the persistence of the changes within the cells occurring during the waiting time and raises the question: What can be measured during this time?

Parameters such as RNA-, DNA-, protein-, carbohydrate- and chlorophyll content did not significantly change nor did the capacity for oxygen evolution. The same result was obtained for respiration, several enzyme activities and polypeptide patterns after SDS- gelelctrophoresis.

The effect of different RNA- and proteinsynthesis inhibitors on the oscillations was investigated. An ethidiumbromide pulse (3 h) (inhibitor of RNA-synthesis) during the waiting time eliminated the release of autospores whereas chlorophyll- and carbohydrate synthesis continued oscillating.

These oscillations also persist if pulses of lincomycine (inhibitor of 70 S translation) were given during the waiting time (Bührmann, 1983).

A pulse of anisomycine (80 S translation inhibitor) eliminates the oscillation in chlorophyll synthesis (fig. 5) whilst allowing carbohydrate production to continue. The addition of cycloheximide during the waiting time eliminates the oscillation of both autospore production and that of chlorophyll. However, a phase shift in the periodicity of carbohydrate production was obtained. (fig. 6).

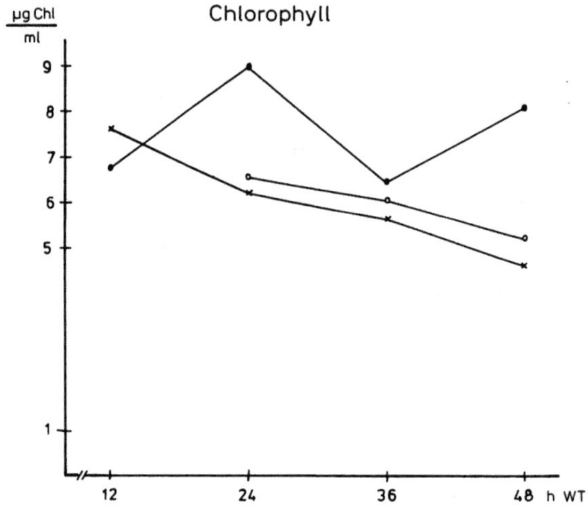

Fig. 5 The circadian oscillation in the capacity of chlorophyll production was eliminated by anisomycine pulses of 12 h: - x -, or 24 h: - o - respectively. The control is given as: - • -.

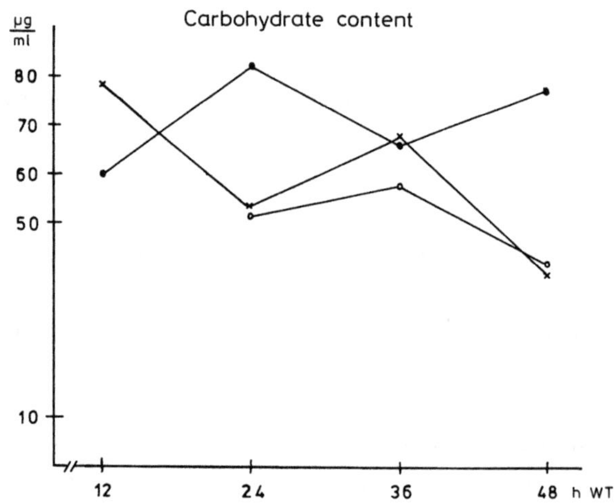

Fig. 6 A phase shift in the circadian oscillation of the carbohydrate productivity was obtained after a cycloheximide pulse of 12 h: - x -, or 24 h: - o - as well. The control data are enclosed as: - • -.

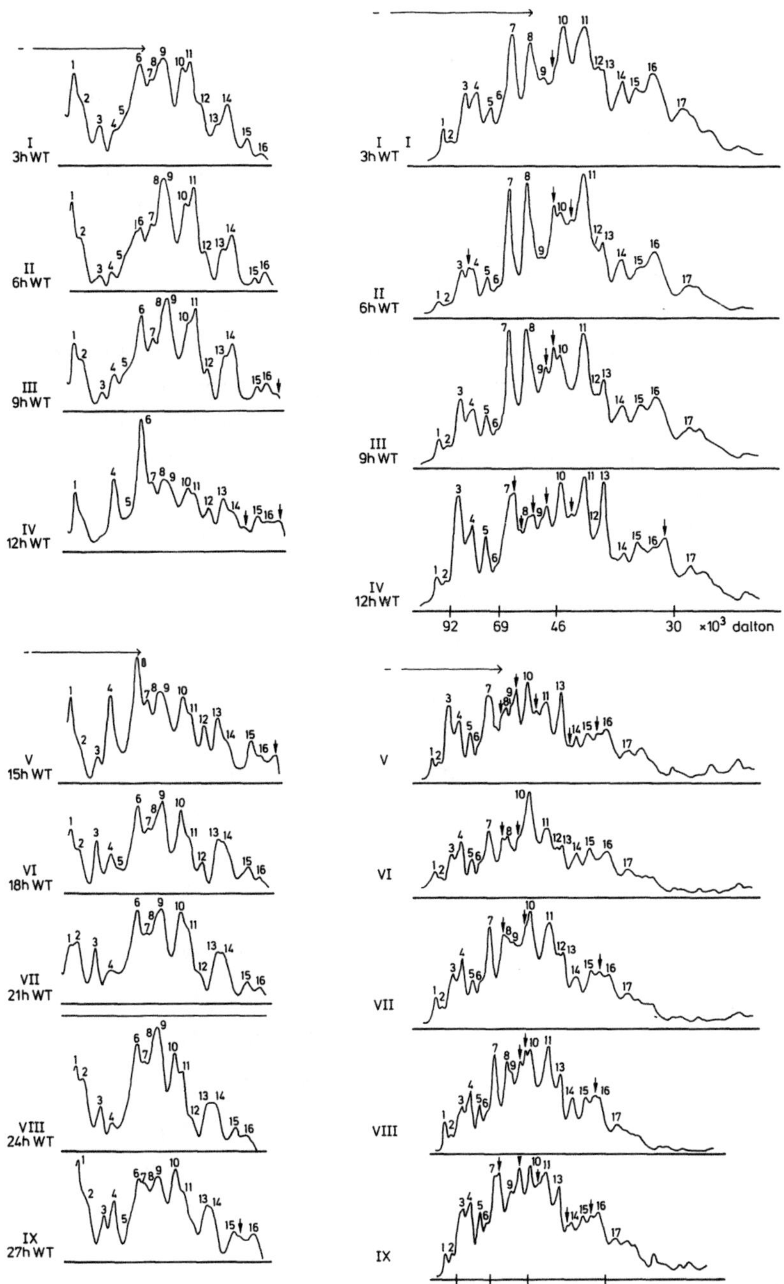

Fig. 7 A circadian oscillation in the ^{35}S-methionine incorporation into protein bands was observed during waiting time. Left part: MW-range less than 25 KD. protein bands 4,13 and 14 show an oscillation. Right part: MW-range more than 25 Kd, protein bands 3,13 and 14 show an oscillation.

In order to get more information on the protein distribution pattern during the waiting time (as already mentioned above no effect on net protein distribution was observed) the proteinsynthesis was evaluated using a radiolabel. To this end ^{35}S-methionine was added for the last 3 h of waiting time. From the result it becomes obvious that there exists a rythm in ^{35}S-methionine incorporation into distinct protein bands (fig. 7 a, b). This was observed for proteins in the range of more and less than 25 KD as well. Even in the presence of cycloheximide some of these changes were observed. (comp. Karakashian and Schweiger, 1976). Summarizing our data: we interpret the effect of anisomycine as decoupling of chlorophyll synthesis and autospore production from the clock. Cycloheximide also has such a decoupling effect for both parameters, but in addition it affects either the clock itself or the coupling mechanism with respect to the carbohydrate production. Changes in the capacity of cell compounds and daughter cell production are based on variations in protein composition as indicated by changes in the protein distribution pattern during waiting time.

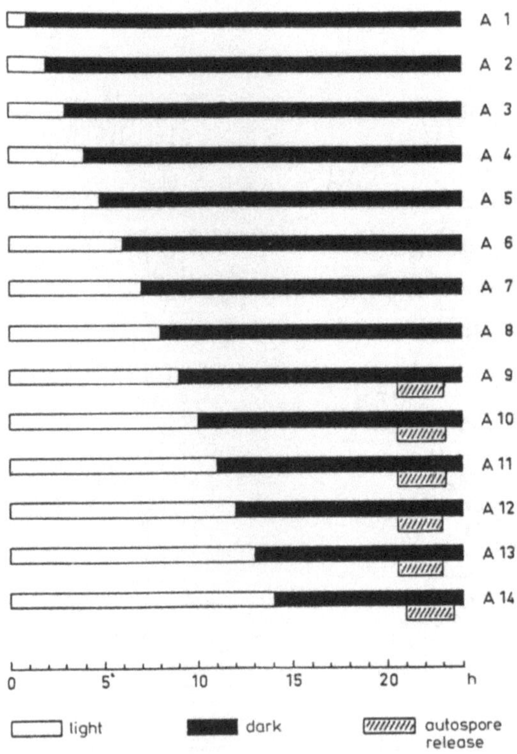

Fig. 8 The release of daughter cells within 24 h was induced after at least 8 h illumination. The release always starts around the 21st h of light dark cycle.

A second phenomenon of time measurement has been described for both Chlamydomonas and Chlorella due to the committment of DNA-doubling and daughter cell number (Mc Ateer et al., 1985). A timing of the release of daughter cells will now be demonstrated for Chlorella cells.
The release of daughter cells within 24 h occurs only if the illumination time exceeds 8 h. A longer illumination increases the number of daughter cells due to a higher production of metabolic intermediates (fig. 8).

However, the release from the mother cell always occurs at the same time, around the 20th - 21st h after onset of the illumination (Lorenzen and Ahlbrodt, 1981). An interruption of the illumination by different dark pulses (fig. 9) resulted in a corresponding delay of the event (e.g. release of autospores). It should be realized that the cells in these experiments were not induced for offspring liberation before the dark pulse.

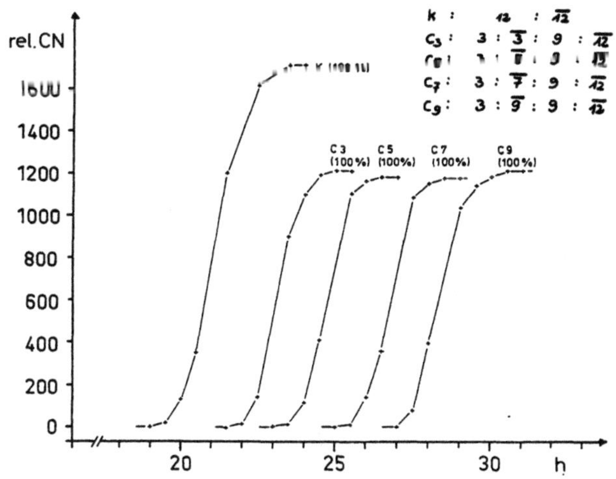

Fig. 9 The timecourse of daughter cell release dependent on different light dark changes as indicated in the upper right hand corner. The bars mean darkness and K stands for control. x-axis: time after onset of illumination, y-axis: daughter cell production (100 = 1.5 x 10^6 cells per ml).

A similar interruption of the illumination after induction of autospore release delayed the event only if the second illumination was longer than 2 h. A second illumination of only 1 h was without any effect on either autospore production or release.

Weak light of different intensities instead of darkness inserted into illumination caused a delay clearly dependent on the light intensities. This indicates the delay to be dependent on the lack of biomass production. It should be said that weak light of about 4000 lux was accepted as a dark treatment.

A further argument based on this interpretation was drawn from the estimation of the carbohydrate contents. These decrease during darkness of the light dark cycle always to a similar value before the release of daughtercells occurs (fig. 10). Such a result was also obtained if the release was delayed due to the dark pulse. Therefore the observed delay is probably due to both: lack in production during the pulse and to metabolic changes (estimated as loss in carbohydrates) during the darkness.

Similar results have been obtained with Chlorella sorokiniana (strain 211-8k). If the illumination was interrupted by dark treatment of 8 h given as two pulses of 4 h each, we got different effects dependent on the light treatment between these pulses. The most interesting result however, was obtained with 5 h light given between the dark treatments (fig. 11).

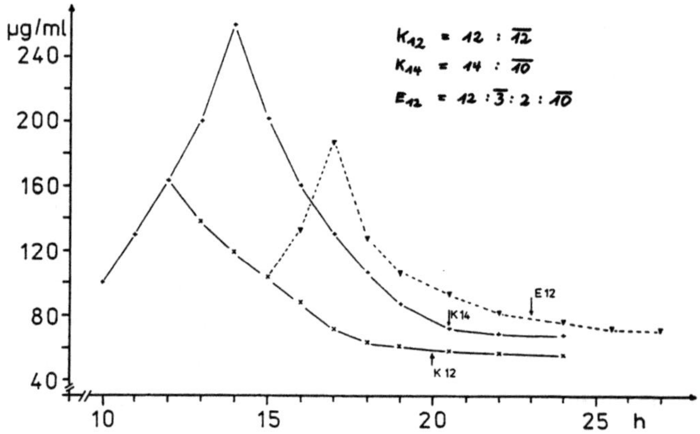

Fig. 10 The timecourse of carbohydrate content (y-axis). Time after onset of the illumination (x-axis). The different light dark cycles employed were as indicated in the upper right hand corner. The arrows indicate the begin of daughter cell release.

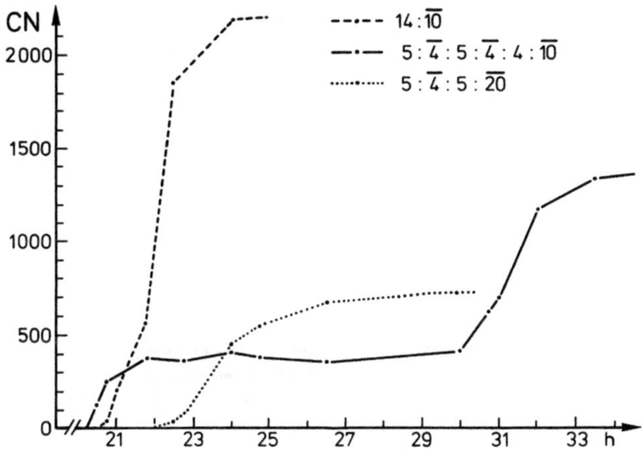

Fig. 11 Two events of daughter cell release were obtained after two separate dark interruptions (4 h each) of illumination. X-axis: time after onset of the illumination, Y-axis: autospore production.

We observed two events; one occured after an 8 h delay (with high yield). Such a splitting of autospore release was also observed if a 6 h dark pulse following a 10 h illumination was used, the second event occured with 6 h delay.
The same behaviour with respect to a delay or splitting of daughter cell release was obtained with Chlorella sorokiniana.
From our results we conclude, that the release of the offsprings in Chlorella growing with a high rate resulting in more than 20 autospores on an average is timed by the onset of illumination which means start of cell material production. This timer seems to be a hourglass type as it has been suggested by McAteer et al. (1985) for both Chlamydomonas and Chlorella dividing by multiple fission. However, we also found that the yield in daughter cell production is timed too. Dependent on the type of light dark change employed the timing of both events may be different.
This is summarized by the following examples (fig. 12):
Remarks to lane 1 and 4 of fig. 12
A dark pulse of 3 h or 4 h after 10 h illumination delayed the release of daughter cells but allows a high production of autospores. The yield therefore was not fixed at the beginning of the second illumination but during the darkness. The induction hence was timed earlier than the yield of autospore production.
Remarks to lane 3 and 5 of fig. 12
The yield of cell production is already fixed if a dark pulse of 8 h or 3 h is given after 10 h or 13 h of illumination respectively before the second

"light on" signal. In the former case the release occurs at the time as that of the control with low yield indicating that the yield has been fixed earlier than the autospore release. In the latter case the cells did not even recognize the 1 h second illumination; no effect on both the yield or the release was observed.

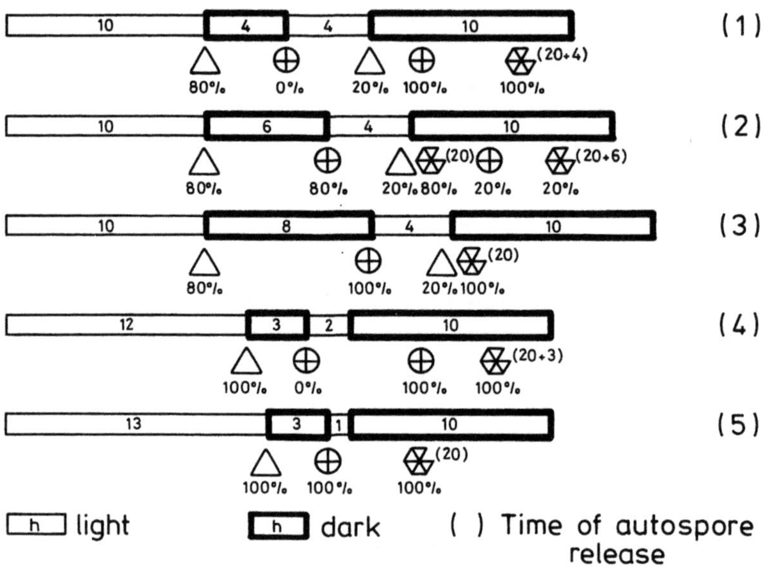

Fig. 12 Committment of the induction for daughter cell release, yield fixation and the time of daughtercell release dependent upon different light dark changes. ( ): time of daughtercell release. For further explanation refer to the text.

Remarks to lane 2 of fig. 12

Some cells were already induced for daughter cell release at the beginning of the dark pulse; their yield of daughter cell production was also fixed. These cells liberate their daughter cells in time with the control with low yield. However, some of the cells (20 %) were induced later (end of the second illumination) and the event occurs with the corresponding delay but with high yield in production.

Literature

Bührmann H (1983) Die Bedeutung von Dauerdunkel für die Produktionsfähigkeit synchroner Chlorella und der Einfluß von Hemmstoffen auf deren circadiane Rhythmik. Dissertation, University of Göttingen

Edmunds L N (1984) In: The microbial cell cycle. ed. P. Nurse, E. Streiblova CRC Press, pp 209-230

Hesse M (1971) Endogene Schwankungen der Produktionsfähigkeit bei Chlorella und ihre Beeinflussung durch Licht sehr niedriger Intensitäten und durch tiefe Temperaturen. Dissertation, University of Göttingen

Hesse M (1974) Z. Pflanzenphysiol. 71, 423-436

Karakashian M W, Schweiger H G (1976) Proc. Natl. Acad. Sci. USA 73, 3216-3219

Lorenzen H, Hesse M (1968) Z. Pflanzenphysiol. 58, 454-456

Lorenzen H, Ahlbrodt (1981) Ber. Deutsch. Bot. Ges. 94, 347-355

McAteer M L, Donnan L, John P C L (1985) New Phytol. 00, 41-58

Wu J T, Tischner, R, Lorenzen H (1986) Plant Physiol. 80, 20-22

DIATOM MITOSIS : IMPLICATIONS OF A MODEL SYSTEM

Jeremy D. Pickett-Heaps
Dept. of Molecular, Cellular and Developmental Biology,
University of Colorado, Boulder, CO 80309

Pennate diatoms have mitotic spindles whose unusually high degree of structural organization allows the changes that accompany mitosis to be analysed in more detail than has been possible with other organisms. The spindles in large organisms are beautiful objects for following mitosis in vivo, while those in smaller species have allowed accurate ultrastructral analysis, for example, by serial section reconstruction. However, these very features render the spindles unusual in appearence while they also display certain other apparently unusual traits (e.g. kinetochore function - see below). As a consequence, objections have been raised that observations on diatom spindles are not representative of the structure and function of more conventional spindles. In this paper, I will outline the major features of the diatom spindle and discuss briefly the extent to which these might be characteristic of all spindles. Wherever possible, reviews will be cited as references.

THE CENTRAL SPINDLE. The pennate diatom spindle contains a prominent "central spindle", clearly visible in live cells (1). This central spindle has been shown from serial section analysis (2,3,4) to consist of two interdigitated half-spindles. With possible exceptions (e.g. 5), it seems likely that most spindles follow this structural arrangement with their continuous fibres organised thus (e.g 6,7,8).

SPINDLE ELONGATION (ANAPHASE B). From the serial section analysis cited above and from in vivo observations, spindle elongation during late anaphase is known to be accomplished by the two half spindles sliding apart (1). Experiments with UV-microbeams (9) and in vitro reactivation of isolated diatom spindles (10) has shown that this sliding is actively generated in the overlap. However, whether similar sliding occurs in other spindles generally is not yet clear. In many animal cells, the central spindle persists as the midbody, often long after mitosis is completed. If sliding does occur in these spindles, it does not go to completion as it does in diatoms in which the half spindles become completely separated (see below).

Algal Development (Molecular and Cellular Aspects)
Ed. W. Wiessner, D.G. Robinson, and R.C. Starr

ARRANGEMENT OF MICROTUBULES IN THE OVERLAP. In some diatoms, the microtubules (MTs) comprising the half spindles, interdigitate in the central overlap so that the MTs are arranged in a square, close-packed array; in addition, any MT from one pole is surrounded by four neighbours from the opposite pole (11). In other diatoms, this general packing arrangement is basically present, with various dislocations and distortions in the array (2,3). Obviously, such a packing arrangment in maximising interaction of MTs from opposite poles, creates the most efficient structural arrangment for a force-generating mechanism which operates by the direct interaction of MTs of opposite structural polarity. The preferential lateral association of MTs from one pole with those of opposite polarity from the other (12), is also apparent in the few other spindle types analysed with the requisite accuracy (6,7,8).

POLARITY OF SPINDLE MTs. The way the diatom spindle assembles suggests that all the MTs from one pole are equivalent; some of them laterally associate with MTs from the other pole to form the overlap described above and thereby collectively create the central spindle while the remainder from each pole penetrate amongst the chromosomes at prometaphase. We concluded (13) that all MTs in each half spindle therefore are of one molecular polarity. This conclusion has since been shown to be true for half spindles of conventional spindles (14,15).

BREAKDOWN OF THE CENTRAL SPINDLE. When anaphase B sliding is completed in diatoms, the half spindles separate and then their constituent MTs disassemble unidirectionally from the overlap end back to the pole (16). UV-microbeam experiments (17) indicate that this disassembly is initiated by the ends of the MTs being freed from the overlap, with disassembly presumably being controlled by their intrinsic polarity. We now believe that this disassembly mechanism provides a primary reason for the sliding of the half spindles that generates anaphase B. Whether such a mode of disassembly occurs in conventional cells is not known. However, if it does, one can explain some aspects of cellular control over MT stability and assembly/disassembly (18) based on this phenomenon.

KINETOCHORE FUNCTION. In diatoms, prometaphase chromosomes in vivo oscillate back and forth along invisible tracks directed at either pole. Very soon, these oscillations cease in each chromosome individually, when the second kinetochore encounters MTs from the other pole. At this point, the second kinetochore stretches in that direction, and the chromosome has then achieved its bipolar, metaphase orientation. Ultrastructural analysis confirms that during the initial oscillations, one kinetochore of each chromosome is sliding along MTs derived from the pole. There is no indication that MTs terminate in these kinetochores

(17). From these observations, we concluded that a primary function of kinetochores is to interact with polar MTs by sliding along them.

This conclusion raises several important issues, for example, whether the kinetochore itself might contain a motor sensitive to the polarity of the MTs it slides over, or whether the motility motor might reside along the polar MTs. Furthermore, this scenario is apparently quite at variance with many observations on conventional cells. In the latter, kinetochore MTs terminate in the kinetochore, and in vitro evidence of MT nucleation at and assembly from kinetochores has been widely reported (summarised in 20). There was formerly almost universal agreement in the interpretation of these and many other observations, that a primary function of the kinetochore is to assemble the MTs of the kinetochore fibre. Subsequently, it became apparent that the polarity of the in vitro nucleation (21) was not that observed in the half spindle (14,15) and at present, the significance of these in vitro assembly observations is uncertain. Now, most workers agree with varying degrees of enthusiasm, that capture of polar MTs by kinetochores, is quite possible. We have proposed that at early prometaphase, conventional kinetochores also slide along polar MTs as in diatoms, before attaching to them; this view is not held by many other workers in the field.

THE SPINDLE MATRIX. Several large diatoms reveal in the electronmicroscope the presence of a matrix, the collar, which permeates the outer MTs of each half spindle, and which runs between the kinetochores and the pole (22). That the kinetochores are attached to this matrix is shown in experiments using colchicine which breaks down the MTs radiating from the poles, while leaving the more stable central spindle intact (23). In this event, prometaphase oscillations cease, while at metaphase, random breakage of either of the two bipolar attachments of each chromosome allows that chromosome to contract elastically to the other pole. Under the electronmicroscope, the resultant spindles have a variable number of chromosomes clustered at each pole, with one of their kinetochores attached to the collar which has now aggregated there. Since in larger diatoms, the bipolarly attached chromosomes normally appear tightly stretched over the central spindle, not surprisingly if the spindle is cut with a UV-microbeam, it instantly collapses (9). From such observations, we concluded that the collar is elastic and that anaphase polar movement of chromosomes is brought about by energy elastically stored in the spindle (22). At metaphase the forces generated in the collar (and stretched chromatin) are balanced over the rigid central spindle and anaphase A is brought about by release of these forces when the chromatids split. The existence of a similar matrix in conventional spindles has not been demonstrated. In our view, there is some indirect evidence that suggests this is so, but the matter is quite open at present.

POLAR AND ANTI-POLAR FORCES. When prometaphase diatom spindles are treated with a variety of metabolic inhibitors, the oscillating chromosomes always move to either pole before stopping movement; when the inhibitors are removed, prometaphase activity soon recommences with the movement of each kinetochore away from the pole (23). We concluded that : i) the movement towards the pole (P-movement) is different to movement away from the pole (AP-movement); and ii) that movement to the pole does not seem to require energy (ATP). We believe that anaphase-A (movement of chromosomes to the pole) is the same as prometaphase P-movement, and therefore is due to energy elastically stored in the spindle matrix (see above). This possibility has been strengthened by observations (Cohn and Pickett-Heaps, in preparation) on the diatom Surirella, in which treatment with dinitrophenol (DNP) at anaphase stops all other cytoplasmic activity, but does not stop chromosome movement to the poles. (Similar experiments with other large diatoms are inconclusive since the kinetochores are stretched almost to the pole at metaphase, and cessation of anaphase A is difficult to demonstrate).

These experiments indicate that there may not be a cellular motor that moves chromosomes to the pole, a possibility with considerable repercussions since all models we know dealing with chromosomal movement, assume that such a motor exists. We then repeated the experiments using metabolic inhibitors on animal (PtK) cells instead of diatoms (24,25). The results were initially puzzling but eventually revealed yet another interesting and significant possibility not taken into account in any previous models of spindle function. First, we found that treatment of PtK cells with the inhibitors caused a contraction of chromosomes around the poles at prometaphase, while the spindle decreased in length. The former observation was exactly what we expected from diatoms, while the latter was apparently a response of the spindle, more pronounced at metaphase, to the increased compression generated on it by bipolarly-attached chromosomes pulling the poles together. The situation at anaphase, however, was more puzzling; anaphase A was immediately and reversibly stopped by the inhibitors. This result is in contrast to what is seen in diatoms. However, the reason for this difference soon became apparent. Since conventional kinetochores have MTs embedded in them, disassembly of these MTs is necessary to permit movement of chromosomes to the pole. We showed in several ways that ATP is probably needed for MT disassembly (24,25). Now it became clear why anaphase stopped in ATP-depleted cells : even if ATP might not be needed for movement of chromosomes to the pole, it is needed for disassembly of the kinetochore MTs that allows this movement to occur.

THE MECHANISM OF ANAPHASE A IN CONVENTIONAL CELLS. The work on metabolic inhibitors started on diatoms, clearly has important implications when extended to conventional cells. We therefore sought means to test some of these possibilities. For a variety of reasons, we had to use permeabilised PtK cells for these experiments (Spurck and Pickett-Heaps, in preparation). When these cells are permeabilised, anaphase stops quite quickly, as endogenous ATP leaks out of the cell. Addition of ATP causes anaphase to restart; a similar result is obtained when anaphase has been stopped by DNP, and the cell permeabilised and then treated with ATP, while maintained in DNP (thus showing that DNP does not directly affect anaphase). We then restarted anaphase similarly using an ATP analogue, ATP- -S which can support phosphorylation but which is not known to reactivate any known cytoplasmic motility system. (We suspect that a phosphorylation reaction requires the ATP needed for anaphase A). Finally, we have been able to reactivate anaphase movement with high Ca^{++} levels or perfusion at 4°C, both treatments known to disassemble MTs. Since it is very difficult to envision how cold treatment might induce the reactivation of a cytoplasmic motor, we conclude that in animal cells too, this motor does not exist, and that polewards movement must come from energy elastically stored in the spindle.

CONTROL OVER MT ASSEMBLY AND DISASSEMBLY. Many of the observations above have been used in a model for explaining how the assembly and disassembly of the different sets of spindles MTs are controlled (16). We believe that the primary function of the kinetochore is to capture polar MTs and later disassemble them in an ATP-dependent manner. We also suggest that a primary factor that determines whether any given MT is shortened or lengthened, is the force (compression or tension) acting on it. In our model, the spindle, whether diatom or conventional, is a self-regulating system in which MT assembly/disassembly is directly affected by the overall forces within it. These forces are generated in the collar or spindle matrix, and by unknown molecular systems operating over the surface of the cytoskeletal MTs.

Acknowledgment. The author acknowledges the grant support (grant # PCM-8309331) from N.S.F. which enabled this work to be carried out.

REFERENCES

1. Pickett-Heaps, J.D., Tippit, D.H. (1978). Cell 14, 455-467.
2. Tippit, D.H., Schultz, D., Pickett-Heaps, J.D. (1978). J. Cell Biol. 79, 737-763.
3. McDonald, K.L., Pickett-Heaps, J.D., McIntosh, J.R., Tippit, D.H. (1977). 74, 377-388.
4. McIntosh, J.R., McDonald, K.L., Edwards, M.K., Ross, B.M. (1979). J. Cell Biol. 83, 428-442.
5. King, S.M., Hyams, J.S., Luba, A. (1983). J. Cell Biol. 94, 341-349.
6. Tippit, D.H., Pillus, L., Pickett-Heaps, J.D. (1980). J. Cell Biol. 87 , 531-545.
7. Tippit, D.H., Fields, C.T., O'Donnell, K.L., McLaughlin, D.J. (1984). Europ. J. Cell Biol. 34, 34-44.
8. McIntosh, J.R., Roos, U.-P., Neighbours, B., McDonald, K.L. (1985). J. Cell Sci. 75, 93-129.
9. Leslie, R.J., Pickett-Heaps, J.D. (1983). J. Cell Biol. 96, 548-561.
10. Cande, W.Z., McDonald, K.L. (1985). Nature 316, 168-170.
11 Tippit, D.H., McDonald, K.L., Pickett-Heaps, J.D. (1975). Cytobiologie 12, 52-73.
12. McDonald, K.L., Edwards, M.K., McIntosh, J.R. (1979). J. Cell Biol. 88, 443-461.
13. Pickett-Heaps, J.D., Tippit, D.H., Leslie, R.J. (1980). Europ. J. Cell Biol. 21, 12-27.
14. Euteneuer, U., McIntosh, J.R. (1982). J. Cell Biol. 89, 338-345.
15. Telzer, B.R., Haimo, L.T. (1981). J. Cell Biol. 89, 373-378.
16. Soranno, T., Pickett-Heaps, J.D. (1982). Europ. J. Cell Biol. 26, 234-243.
17. Leslie, R.J., Pickett-Heaps, J.D. (1984). Cell 36, 717-727.
18. Pickett-Heaps, J.D., Tippit, D.H., Cohn, S.A., Spurck, T.P. (1986). J. Theoretic. Biol. 118, 153-169.
19. Tippit, D.H., Pickett-Heaps, J.D., Leslie, R.J. (1980). J. Cell Biol. 86, 406-412.
20. Pickett-Heaps, J.D., Tippit, D.H., Porter, K.R. (1982). Cell 29, 729-744.
21. Bergen, L.R., Kuriyama, R., Borisy, G.G. (1980). J. Cell Biol. 84, 151-159.
22. Pickett-Heaps, J.D., Spurck, T.P., Tippit, D.H. (1984). J. Cell Biol. 99, 137s-143s.
23. Pickett-Heaps, J.D., Spurck, T.P. (1982). Europ. J. Cell Biol. 28, 83-91.
24. Spurck, T.P., Pickett-Heaps, J.D., Klymkowsky, M.W. (1986). Protoplasma 131, 47-59.
25. Spurck, T.P., Pickett-Heaps, J.D., Klymkowsky, M.W. (1986). Protoplasma 131, 60-74.

WALL MORPHOGENESIS IN CENTRIC DIATOMS

Anna-Maria M.Schmid
Universität Salzburg, Botanik
Hellbrunnerstr.34, A-5020 Salzburg

Diatoms are enclosed within a rigid cell wall which consists of two thecae overlapping at the girdle like a bipartite box. Cell division occurs by binary fission parallel to the valvar plane. Each daughter cell inherits one half of the mother cell wall; each half in both cells then becomes the epitheca, while each new hypotheca is created within an intracellular membrane bound compartment, the "silica deposition vesicle" (SDV) (Figs.1,3) (Drum et al.1964; Reimann 1964). This sequence of events causes a decrease in size during successive divisions which at a certain specific stage is compensated for by sexual reproduction which generates a huge auxospore, within which the initial valves of a new clone are laid down. The initial valves are ± hemispherical; the species specific surface form of the vegetative valves is created with the first division of the initial cell.

Because of differences in their sexual reproduction and overall symmetry, diatoms are divided into two major groups: the oogamous Centrales, with mainly tri- to omniradiate, but also bipolar symmetry, and the isogamous Pennales with bipolar, and only as an aberration (Schmid 1985) tripolar symmetry. This chapter will deal with centric diatoms only; for references on Pennales see Schmid et al.(1981), Pickett-Heaps (1983), Edgar et al.(1984a).

The enormous variety and complexity of shapes and patterns displayed by the valves stimulates the question of "how" and "why" they are formed in such a way: what determines their outline, their topography and their pattern. Early LM-observations showed that the formation of the new valve is a growth process initiated at the pattern center, which in centrics is either a ring-like structure, the annulus (Stosch et al. 1973), or a central area (Schmid et al.1983); then the growth occurs centrifugally outwards to the periphery. Several authors have claimed that the primary silicification site in centric diatoms is the "labiate process", an opening in the valve, with a scapuliform internal structure and often differentiated into an external tube (e.g. Li et al.1985a,b). However, the labiate process (LP) coincides with the primary silicification site only when it is situated within the annulus. The original observation of Stosch et al. (1973) has been confirmed with TEM and SEM studies (Schmid et al.1979,1983, Schmid 1984c,1986a).

The main characteristics of valve formation in Thalassiosira eccentrica and Coscinodiscus wailesii are briefly outlined below and illustrate the complexity of events that can be involved in valve morphogenesis: in both cases, the construction element of the valves is a very common and possibly very archaic structure, a loculate areola, a small chamber, hexagonal in outline and covered with a top and a base wall. These limiting layers are either perforated like sieves or open with a round hole, the foramen.

Algal Development (Molecular and Cellular Aspects)
Ed. W. Wiessner, D.G. Robinson, and R.C. Starr

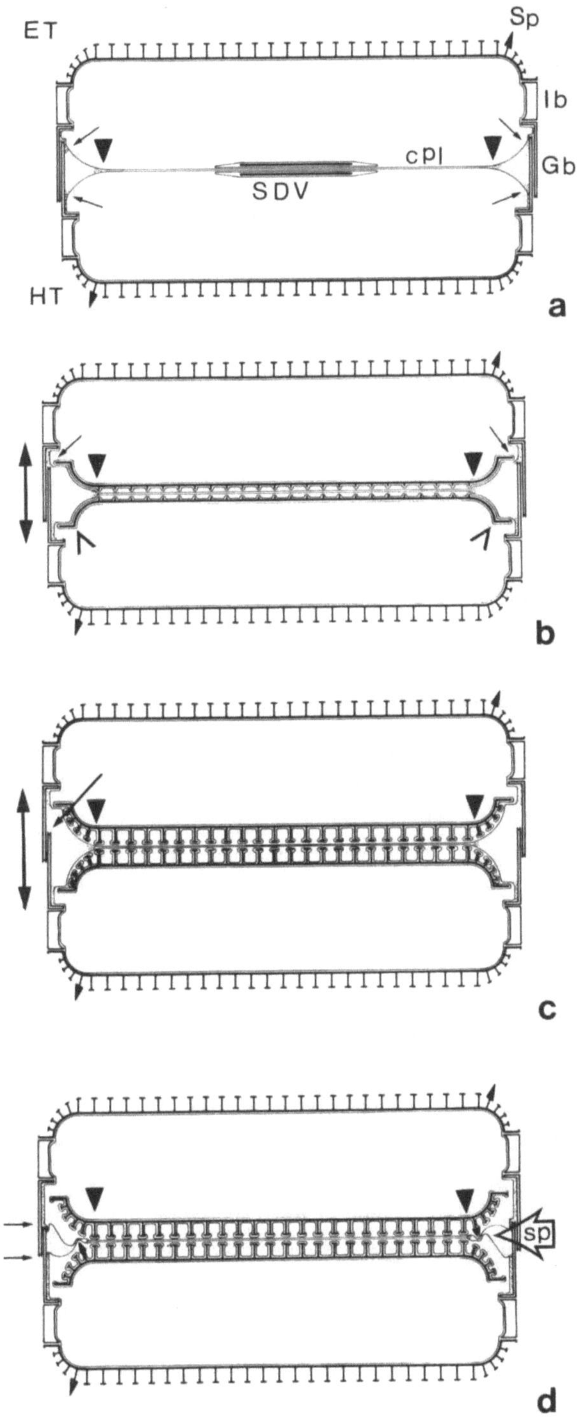

Fig.1 Schematic drawing of valve formation in Thalassiosira eccentrica.

a) Base layer formation: siblings are circumferentially adhered (arrow heads) until valve release; both cells adhere to the girdle bands (Gb) of the hypotheca (HT) (arrows); rounded cPL at the margin between the adhesion sites due to surface tension; SDV distally adhered to the cPL. Sp=spines; ET=epitheca; Ib=intercalary band.

b) Formation of chamber side walls: addition of material at the side facing the cPL; delay in the marginal region; open arrow indicates pronounced angle at the valve edge; thecae start to slide apart (double arrow); both cells still adhered to the HT-Gb.

c) Outer layer formation: thecae are further apart (double arrow); cell in the ET disconnected from the HT-Gb (arrow).

d) Final moulding and stabilisation: cPL "crawls" along Gbs (small arrows) for new spot like cell-cell contacts which become the molds for the spines (open arrow).

Thalassiosira has been taxonomically separated from Coscinodiscus (besides other characteristics) because of its "inverted" architecture: sieve plates are located at the distal side in Coscinodiscus (Fig.3e,f) and at the proximal side in Thalassiosira. Regardless of the position of foramen and sieve in their chambers, their growth sequence is the same: the initial horizontal differentiation is a centrifugal growth of the proximal or base layer. A vertical differentiation of the chamber side walls follows, and finally the distal or outer layer of the valve is formed (Figs.1,3).

The first sign of a developing valve as seen in the TEM is a minute SDV closely apposed to the plasmalemma (Fig.1); the SDV then expands centrifugally by branching radial tubes; the tubes become reticulate by putting out crossextensions (Fig.3a) (Schmid et al.1979,1981,1983). Common to both the currently investigated centric and pennate species, is the dependency of the primary silicification site on the nuclear position at this stage. The nucleus migrates into the girdle region for performing mitosis. If its later migration back to the center of the newly created surface at the cleavage furrow is inhibited arteficially, the pattern center of the new valve is displaced (Schmid 1979). Normally the new valves start to be formed while the nuclei are still in an excentric position (Schnepf et al.1980, Li et al.1985a, Schmid unpubl.), then they concurrently migrate to the center. Acytokinetic mitosis resulting in a binucleate cell, gives rise to the formation of a single valve bearing 2 pattern centers (Schmid 1984b,and unpubl.). The functional implication of the nuclear position is yet to be determined.

In Thalassiosira the growth of the SDV in all developmental stages is the consequence of addition of small vesicles whose origin was not clear at that time (Schmid et al. 1979). Growth of the SDV by vesicular addition was observed for a number of centric diatoms (Schnepf et al.1980; Li et al.1985a,b), whereas in pennates the situation is not clear yet. In Coscinodiscus the vesicles were seen to be Golgi-derived (Schmid 1984c,1986a).

For shaping a diatom valve we can assume that silicon, the major building material, is transported and moulded while in aqueous solution, and it precipitates and hardens by polycondensation into opaline silica. Silica is amorphous without any inherent form. Like concrete, it displays grossform only when moulded from the outside. The silicalemma as a naked biomembrane is however, not rigid; therefore, the SDV does not have moulding capacity. Thus the SDV has to be moulded itself by the external application of either pressure or tensile forces. The moulding process for a new valve is a precisely timed sequential interaction between the plasmalemma created during cleavage (cPL), membraneous and cytoskeletal organelles, turgor pressure and the mother cell wall.

1) Outline of the new valve

Since the vegetative valves are formed when the sibling cells are still enclosed by the mother cell wall, their outline is predetermined by the outline of the latter (Schmid 1985,1986b). Thus the issue how the outline is controlled is reduced to what occurs during the initial valve formation: a zygote (in centric diatoms) swells into a "hydrosphere", probably as a consequence of disintegration of the cytoskeleton which attend meiotic division (Schmid 1984b). A layer of small siliceous scales are formed subjacent to the organic envelope and then provide some degree of external reinforcement. Circumferential adhesion of the protoplast to the auxospore cover in the prospective cell equator, together with a slight contraction of the cell then determine shape and outline of the initial valves. All forms deviating from a simple circular outline need a moulding "perizonium", as is present in pennates, within which the more complex shape arises (ref. in Stosch 1982).

2) The moulding surface for the topography of the new valve

Three types of plasmamembranes are distinguishable in Coscinodiscus: I) the plasmalemma (and fibrous lamina -fLa) created during cleavage (cPL) whose main function seems to be that of moulding the siliceous part of the valve (Schmid 1984c,1986a,b,c); II) the plasmalemma created upon completion of the siliceous part of the cell wall (sPL) forming the organic layer that seals the siliceous chambers from the cytoplasm (Fig.2)(Schmid 1984c,1986a,b,c); III) the metabolically active plasmalemma (mPL) which is ruffled into a distinctive pattern of microplicae and is clearly separated from those cytoplasmic surfaces involved in morphogenesis (Schmid 1986c) (Fig.2).

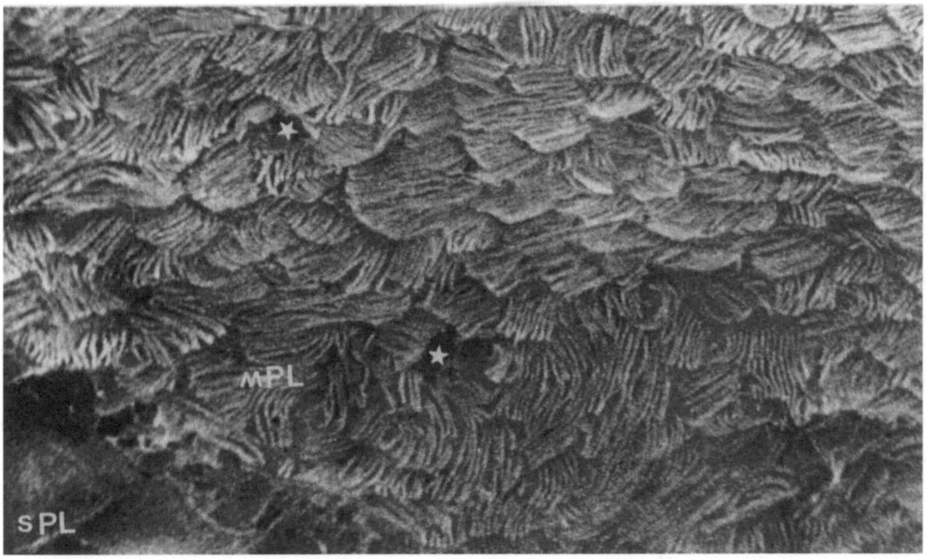

Fig.2. Coscinodiscus wailesii: plasmalemma moulding the organic layer of the girdle (sPL), delimited at the valve edge from the metabolically active PL (mPL); torn during valve liftening along the transient adhesion zone. GA/Os/CPD, SEM x 8.000.

The shape of the cleavage furrow in many instances directly determines the shape of the future valve; this shape depends on the turgor, the tension of the new plasmatic layer, and whether the mother cell adheres to either all of its girdle bands prior to cleavage (Coscinodiscus), or only to the last and the first ones (Thalassiosira). In a number of cases the cleavage furrow, which after cytokinesis adheres both siblings to the hypotheca (Fig.1a,b) (Schmid 1986c), is premoulded prior to or during valve deposition, either by postcytokinetic swelling (Stosch et al.1964), or by the nucleus (Stosch et al.1973), or by adhesion and local contractions, which could involve secretion of fluids which resist compression (ref.in Schmid 1984a). Contact between the sibling surfaces causes their shape to be mutually influenced. If adhesion is circumferential at the margin, then the mutual effects depend upon the tension of the circumscribed area: high tension creates a flat surface as in Thalassiosira (Fig.1), low tension causes complementary undulations or ridges and grooves (ref.in Schmid 1984a). The 2 cells then behave as one entity, a 2-celled tissue, during morphogenesis.

In T.eccentrica both siblings not only adhere to the girdle bands of the mother cell wall, but also to each other during valve formation. As result of these adhesions, tension seems to be developed across the surface of the forming valve and this tension appears to deform the chamber pattern from its radial in the base layer to a more tangential pattern at the distal side of the valve. Marginal spines, which are formed later in morphogenesis seem to reflect these tangential lines of tension. Up to the formation of the outer layer of the valve the plasmamembranes of the sister cells are aligned with the forming valves, being the moulding surface for the outer layer (Fig.1); then the cPL grows and folds out, using the girdle bands of the mother cell wall as substrate for what appears in life cells as "crawling motion" to make new spotlike cell-cell-contacts at the margin; this event is accompanied by a pervalvar shift of the interconnected surfaces of the siblings. As soon as the new cPL-contacts are made, the cPL is retracted and, owing to the tensile stress tent-like lifted at the new connecting points. Final silicification occurs while the cPL is in this stretched state, providing optimal stability of the valve (Fig.1d) (Schmid 1984a,c).

3) Moulding of the valve pattern

In C.wailesii a very complex moulding machinery acts in a precisely timed sequence of interactions (Schmid 1984c,1986a). A prepattern is already present on the cleavage furrow prior to the appearance of the SDV: ER-cisternae outline the future SDV location of the base layer, and the shape and position of the future areolae are predetermined by large areolar vesicles (AV) tightly bound to the cPL. Their non-random distribution and their attachment indicate that the primary prepattern for the Coscinodiscus valve is located in the plasmalemma and its cytoplasmic coat, the fibrous lamina, and is created during or shortly after cleavage. The SDV grows centrifugally from the boundary of the central area as a tubular membrane system in the distal gaps between the AVs. The ER which had foreshadowed the SDV location, gives way to the outgrowing tubes and attaches to their proximal sides, where it remains until valve release. At the growing tips of the tubular SDV, fibrous dense domains possibly provide the pulling forces for the horizontal expansion of the SDV. Similar dense domains appear to be involved in labiate process formation of this species. The SDV has to bend around the areolar vesicles during outgrowth, thus forming rings, the prospective foramina of the future

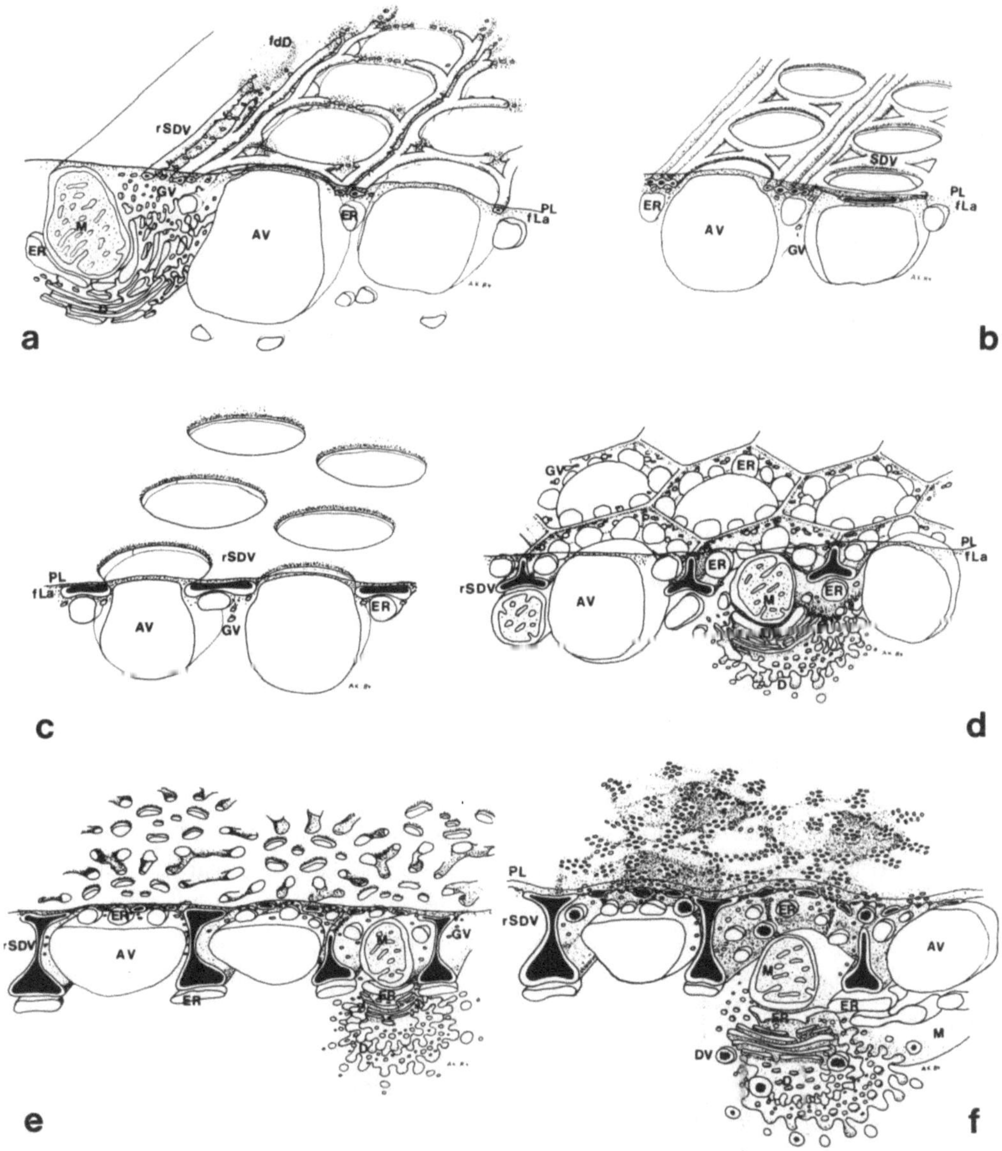

Fig.3. Coscinodiscus wailesii: schematic drawing of valve development. a - c) development of the base layer; d) development of the chamber side walls; e - f) development of the outer layer. fLa=fibrous lamina; fdD=fibrous dense domain; D=dictyosome; GV=Golgi vesicles; AV=areolae vesicles; ER=endoplasmic reticulum; M=mitochondrium; rSDV=reticulate SDV; PL=plasmalemma. (from Schmid 1984c).

chambers. The SDV is attached to the cPL by the fibrous lamina and is sandwiched vertically between cPL and ER, and laterally bordered by the AVs (Fig.3a-c).

Vertical differentiation (formation of the chamber side walls) follows partial detachment of the SDV from the plasmalemma and the subsequent immigration of ER into this area, which may inhibit lateral expansion of the SDV and thus cause vertical growth. The increased vesicle volume distal to the foramina may exert even lateral compressive stress and force the vertically growing walls into hexagonal arrangement (fig.3d). Vertical growth is terminated by detachment of the areolar vesicles from the cPL, which allows the SDV to expand again laterally and to adhere to the cPL; premature detachment of the areolar vesicles causes the outer layer to sit directly on the base layer (unpubl.). At the same time ER-vesicles collect distally to the AVs and mould the pores of the outer layer (Fig.3e). The ER then detaches and finally the pores are filled with a very delicate lattice whose meshsize has the same dimensions as those of distal Golgi cisternae (Fig.3f) (Schmid 1984c,1986a). The role of the ER is not only similar to that in chrysophyte scale morphogenesis (Schnepf et al.1969), it also resembles the situation found in pollen wall formation (ref.in Dickinson 1970). The role of the ER can be partly taken over by mitochondria (Schmid 1986a,b), which also function in shaping raphe canals (ref.in Edgar et al.1984b).

Specific cytoplasmic fibrous domains, the "labiate process apparatus" and "struttet process apparatus" are involved in the formation of labiate process (Schnepf et al.1980; Li et al.1985a; Schmid unpubl.) and the strutted process (Fig.4 by D.Schulz in Schmid 1986b). An analogous structure in pennates is the "raphe fiber"(Edgar et al.1984b). Labiate and strutted processes are the attachment sites of the cell to the cell wall (Schmid 1986c), and at the same time openings to the environment. The raphe fiber in pennates was shown to be stabilized by microtubules (ref.in Edgar et al.1984a,b). Microtubules have not yet been found in direct connection with the LPA and SPA. The sensitivity of their distribution and structure to MT-inhibitors suggest however, that MTs are involved. Unlike the situation in several pennates (Pickett-Heaps et al.1979;Edgar et al.1984a) no microtubule center (MC) is present during the formation of the base layer in C.wailesii; as soon however, as the first layout of the valve patterning has reached the future valve edge, the MC can be found as a dense structure close to the nucleus with the spindle precursor for the next division sandwiched between them. This sequence of events suggests that the MC-spindle-precursor-complex is involved in the formation and distribution of labiate and strutted processes, the future attachment sites of the protoplast to its cell wall. Anti-MT drugs cause clustering and malformation of labiate and strutted processes (Schmid 1984b, and unpubl.).

4) The organic part of the valve and valve release

As soon as the siliceous part of the valve in C.wailesii is completely moulded, the Golgi-apparat of the entire cell switches to produce a new type of vesicles with dark content. The vesicles fuse with the proximal silicalemma and discharge their content into the SDV, thus contributing organic coat material. This coat substance could provide the stop mechanism for further silica deposition either by masking the precipitation sites or by changing the milieu which is required for deposition. The up to now reticulate SDV becomes by this membrane addition a continuous sac, which enables the proximal silicalemma to retract from the sieve plate and the cytoplasm to retract from the chambers. The distal silicalemma fuses with the cPL above the valve center in a reticulate way and the membranes seem to be pulled back over the valve edge. This membrane behaviour indicates that valve liberation in diatoms is a normal exocytotic process. The membrane incorporation of the dense vesicles probably causes a gradual membrane change, that is, the proximal silicalemma gradually becomes the new plasmalemma. It now spans across the foramina; the vesicles continue to fuse and the organic content is distributed along the stretched membrane, thus forming the organic layer that seals the chambers from the cytoplasm (Schmid 1986a).

Valve formation in diatoms is a delicately balanced feed back controlled ontogenetic system which involves almost all cell organelles and where the sequence of events is self propagating as long as it is not disturbed. But there is no information on the feedback circuit and how the information is carried on.

References

Dickinson,H.G. (1970) Cytobiology 1, 437-449.
Drum,R.W.,H.S.Pankratz (1964) J.Ultrastruct.Res.10, 217-223.
Edgar,L.A.,J.D.Pickett-Heaps (1984a) J.Phycol.20, 47-61.
Edgar,L.A.,J.D.Pickett-Heaps (1984b) Progr.Phycol.Res.3, 47-88.
Li,C.W.,B.E.Volcani (1985a) Protoplasma 124, 10-29.
Li,C.W.,B.E.Volcani (1985b) Protoplasma 124, 147-156.
Pickett-Heaps,J.D. (1983) J.Phycol.19, 269-281.
Pickett-Heaps,J.D.,D.H.Tippit,J.A.Andreozzi (1979) Biol.Cellulaire 35,199-206.
Reimann,B.E.F. (1964) Exp.Cell Res.34, 605-608.
Schmid,A.M. (1979) Europ.J.Cell Biol.20, 125.
Schmid,A.M. (1984a) Proc.7th Internat.Diatom Symp. ed.D.G.Mann, pp.71-95.
Schmid,A.M. (1984b) Proc.7th Internat.Diatom Symp. ed.D.G.Mann, pp.47-70.
Schmid,A.M. (1984c) Communications of Inst.Lightweight Structures 28, Stuttgt.pp.300-317.
Schmid,A.M. (1985) In:Contributions to the Palaeolimnol. of the Trumer Lakes and the Lakes Mondsee, Attersee and Traunsee. ed.D.Danielopol,R.Schmidt, E.Schultze; Österr.Akademie Wiss., pp.65-78.
Schmid,A.M. (1986a) Proc.8th Internat.Diatom Symp., ed.M.Ricard, in press
Schmid,A.M. (1986b) Proc.8th Internat.Diatom Symp., ed.M.Ricard, in press.
Schmid,A.M. (1986c) In: Cytomechanics. ed.J.Bereiter-Hahn,O.R.Anderson,W.E.Reif, in press.
Schmid,A.M.,D.Schulz (1979) Protoplasma 100, 267-288.
Schmid,A.M.,M.A.Borowitzka,B.E.Volcani (1981) In: Cytomorphogenesis in plants. ed.O. Kiermayer, Springer Wien,New York, pp.63-97.
Schmid,A.M.,B.E.Volcani (1983) J.Phycol.19, 387-402.
Schnepf,E.,G.Deichgräber (1969) Protoplasma 68, 85-106.
Schnepf,E.,G.Deichgräber,G.Drebes (1980) Pl.Syst.Evol.135, 265-277.
Stosch,H.A.von (1982) Bacillaria 5, 127-156.
Stosch,H.A.von,G.Drebes (1964) Helgol.Wiss.Meeresunters.11, 209-257.
Stosch,H.A.von,G.Theil,K.V.Kowallik (1973) Helgol.Wiss.Meeresunters.25, 384-445.

CELL ELONGATION IN THE RED ALGA GRIFFITHSIA: CONTROL BY LIGHT, IONS, AND AN ENDOGENOUS GLYCOPROTEIN HORMONE

Susan D. Waaland
Department of Biology
University of Puget Sound
Tacoma, WA 98416 U.S.A.

In cells which elongate by tip growth, wall loosening and wall synthesis are localized at the apex of the growing cell. This localization of biochemical and developmental activities is reflected in an accumulation of specific organelles and of sites of exocytosis at the growing point (Sievers and Schnepf, 1981; Gooday, 1983). We are studying the control of cytoplasmic localization and localized growth in the red alga Griffithsia pacifica Kylin (Rhodophyta, Ceramiales). In this alga, there are two types of cells that elongate by tip growth: ordinary rhizoid apical cells and specialized repair shoot cells. The growing apex of both types of cells has a characteristic organization. The apical dome contains a highly refractile zone; behind the refractile zone is a deeply pigmented chloroplast band; basal to this band, the cytoplasm is nearly colorless. Ultrastructural examination of growing tips of rhizoids and repair shoot cells reveals that the refractile zone contains nuclei, endoplasmic reticulum, ribosomes and numerous active dictyosomes. Each dictyosome has a mitochondrion on its forming face. The cisternae of the dictyosomes are inflated. Large, dictyosome-derived vesicles can be seen to fuse with the plasmalemma in the apical elongation zone. This type of organization is similar to that found in tip-growing cells in other organisms (Sievers and Schnepf, 1981).

Hormonal Control of Repair Shoot Elongation: While the morphology and pattern of growth of rhizoids and repair shoot cells are similar, these cell types differ in their physiology and in their requirements for sustained elongation. The repair shoot cell is a specialized cell produced during cell repair (Waaland and Cleland, 1974). Elongation in a repair shoot cell is regulated by an endogenous, glycoprotein hormone, "rhodomorphin". Rhodomorphin, from G. pacifica , is a species-specific, acidic glycoprotein with a molecular weight of 14-17,500. It is produced by rhizoid cells and

Algal Development (Molecular and Cellular Aspects)
Ed. W. Wiessner, D.G. Robinson, and R.C. Starr

regulates the formation, growth and morphogenesis of repair shoot cells (Watson and Waaland, 1983; 1986). In the presence of rhodomorphin, repair shoot cells have a prominant apical refractile zone and elongate by tip growth. When the hormone is withdrawn, repair shoot cell elongation continues for 1.5-2 hr during which time, the refractile zone gets progressively smaller and finally disappears. Elongation then ceases and the cell widens laterally reverting to a normal shoot cell pattern of growth and morphogenesis (Waaland and Lucas, 1984). If hormone is readministered within 8 hrs after its withdrawal, the refractile zone reforms at the repair shoot apex and elongation restarts (Waaland, 1986).

Photocontrol of Rhizoid Elongation: Rhizoid elongation is not dependent on the presence of rhodomorphin; however it can be altered by changing light conditions. First the direction of elongation may be altered by unilateral light (Waaland et al., 1977). Second, rhizoids require light for sustained elongation. When rhizoids are placed in continuous dark, their elongation continues for 24-30 hrs, then ceases. This cessation of growth is accompanied by changes in the morphology of the rhizoid tip. The dark-grown tip loses its characteristic refractile zone and the chloroplast band disperses. On the ultrastructural level, there is no longer a concentration of dictysomes or vesicles at the rhizoid apex; in addition, chloroplasts are no longer excluded from the apex and are found in the apical zone as well as along the sides of the cell. The dictysomes, still with associated mitochondria, have flattened cisternae and appear to be inactive. If a dark-treated plant is returned to the light, the rhizoid tip reorganizes. The refractile zone and chloroplast band are reestablished and growth restarts.

Thus apices of growing rhizoids and repair shoot cells have a characteristic cytoplasmic organization typified by localized accumulation of specific organelles and a localized zone of vesicle fusion with the tip. Dark-treated rhizoids and hormone-deprived repair shoot cells lose this organization and cease to elongate. Because cytoplasmic organization and localized growth can be manipulated in these cells, one can investigate the role of various internal factors in the regulation of localized organelle movements and localized exocytosis. We have investigated the role of trans cel-lular currents, and of specific ions in elongation of *Griffithsia* cells.

Transcellular Currents: In many cells which elongate by tip growth, a role for transcellular ion currents in the establishment and maintenance of polarized growth has been postulated. Regions of localized inflowing current are found at the tips of many of these cells; in some cases localized inflowing currents preceed the initiation of localized growth (Nuccitelli, 1983; 1986). The role of endogeneous transcellular ion currents in regulating elongation and maintaining localized growth and organelle localization in Griffithsia has been studied (Waaland and Lucas, 1984). Inflowing currents are localized at the tips of growing rhizoid and repair shoot cells. However these currents continue to be localized at the repair shoot tip even when elongation ceases following rhodo-mor-phin withdrawal. In addition, while these currents disappear in dark-treated rhizoids, they are not reestablished preceding reorganization of the tip and reinitiation of growth when the rhizoid is returned to the light. Thus localized inflowing currents do not appear to be causal in the maintenance or reestablishment of localized organelle accumulation and localized elongation in Griffithsia rhizoids or repair shoot cells.

Role of K^+ in Elongation: While transcellular current per se does not seem to be responsible for maintenance of elongation in tip-growing cells of Griffithsia, it is likely that influx and locally high concentrations of specific ions do play an important role. Therefore we have begun to characterize the role of specific ions in regulating growth in these cells. Findlay et al. (1969) have shown that cells of several species of Griffithsia actively accumulate K^+ in their vacuoles to levels approximately 50 fold that found in the medium. We have investigated the role of K^+ in maintaining rhizoid elongation. Plants, grown in enriched sea-water, were transfered to a defined medium containing K^+ concentrations of 0.0, 1.0 mM, 3.0 mM, 10 mM, and 100 mM. Rhizoid elongation and morphology was monitored. In 100 mM K^+, rhizoids lysed at their apices. In some cases, the tip resealed and elongation restarted. The effects of the other concentrations of K^+ on elongation are shown in Figure 1. In the absence of K^+ elongation

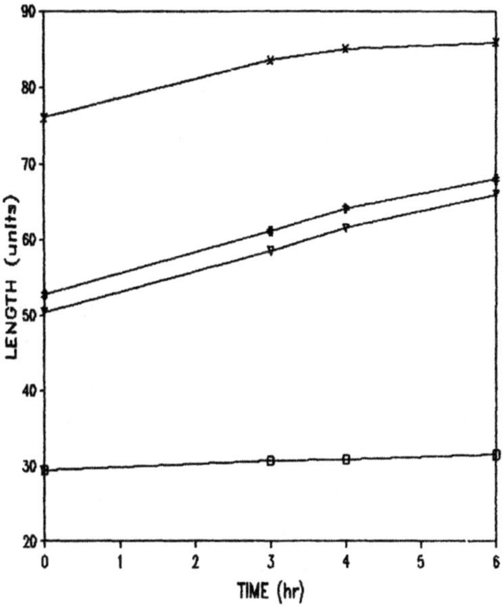

Figure 1: Effect of [K^+] on Rhizoid Apical Cell Elongation. Plants were started from single cells and cultured for 7 days as previously described (Waaland and Watson, 1980). For these experiments, a defined medium (Nuccitelli, 1978) was used. Osmotic concentration was kept constant by varying [NaCl] with different [KCl]. (-□- 0.0 mM K^+, -X- 1.0 mM K^+, -∇-3.0 mM K^+ -#-10.0 mM K^+ (Control))

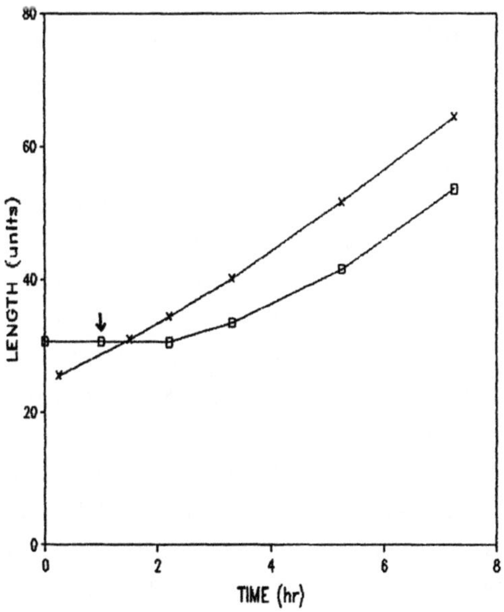

Figure 2: Effect of Monensin on Rhizoid Apical Cell Elongation. Plants were started from single cells and cultured for 7 days as previously described (Waaland and Watson, 1980). Medium containing monensin was prepared by diluting a stock solution of 5 mM monensin in 95% ethanol 1:500 in enriched seawater. (-□- 10 µM monensin, -X- Control) Monensin was added at time=0 and removed at time=1 hr (arrow)

ceased immediately. In 3 mM K^+ rhizoid elongation continued at the control (10 mM K^+) rate, while 1 mM K^+ rhizoids elongated for several hours and then stopped. In the absence of external K^+, the refractile zone at the rhizoid apex gradually became smaller. This change took several hours even though elongation stopped immediately.

The K^+ levels inside cells were altered in a second way. Plants were treated with the K^+ ionophore, valinomycin an ionophore which allows K^+ to diffuse freely across membranes thus collapsing gradients in [K^+]. When plants were treated with 5 uM valinomycin, rhizoid elongation did not stop completely however its rate of elongation slowed to one quarter that of the control. Therefore although gradients in [K^+] are collapsed by valinomycin, localized elongation is maintained. These results suggest that K^+ is required to maintain elongation of rhizoids; however a K^+ gradient is not required to sustain elongation. Bisson and Kirst (1979) have shown that K^+ acts as an osmotic solute in *Griffithsia monilis*. It is likely that this is the role that it plays in maintaining elongation in *G. pacifica*. K^+ does not appear to play a direct role in maintaining cytoplasmic stratification in these cells.

Effect of Monensin and Nigericin: The effect of two additional ionophores, nigericin and monensin, on rhizoid organization and elongation has been investigated. These monovalent ionophores catalyze the diffusional exchange of monovalent cations (K^+, Na^+, H^+). Nigericin shows a higher affinity for K^+ and monensin shows a higher affinity for Na^+. When rhizoids or repair shoot cells of *Griffithsia* were treated with either 10 μM monensin or 5 μM nigericin, elongation of both types of cells was inhibited. If the ionophores were washed out within 24 hrs, elongation resumed. The kinetics of the rhizoid response to monensin are shown in Fig. 2. Monensin was added at time zero and rhizoid elongation stopped immediately. When the ionophore was washed out after 1 hr, rhizoids resumed elongation witha lag of 2-4 hrs. While the effects of short-term treatment with monensin and nigericin are reversible, extended exposure of plants to these ionophores results in cell death. Normal apical shoot cells are most sensitive to these ionophores and die within 4-8 hours of treatment. Apical rhizoid cells and repair shoot cells survive in nigericin and monensin for at least 30 hours.

The effects of monensin and nigericin on rhizoid morphology are quite different than those seen after either treatment with valinomycin or removal of K^+. In the presence of either ionophore, the refractile zone of the apices of either rhizoids or repair shoot cells increases in size even though elongation is inhibited. At the light microscope level, the texture of this zone appears to change; there appear to be large vesicles in the tip region of the cells. In addition to their effects on rhizoids and repair shoot cells monensin or nigericin induce a striking change in the morphology of normal shoot cells. In these cells, chloroplasts, which are normally evenly distributed along cells, accumulate at the tops of intercalary cells. This response is not observed with either valinomycin treatment nor K^+ removal from the medium.

We have examined the ultrastructure of rhizoid apical cells which have been treated with 10 µM monensin for various times. Within 10 min after the start of monensin, the trans-cisternae of the dictysomes become very swollen. At 30 min the dramatic swelling has increased. Cis-cisternae do not appear to swell. Vesicle fusion with the plasmalemma does not appear to occur. The ultrastructure of of other organelles (mitochondria, nuclei, etc.) does not appear to be affected by monensin. The effects of monensin on dictysomes of Griffithsia rhizoids are similiar to those observed in a number of animal and plant systems (Robinson, 1981; Mollenhauer et al., 1983; Ledger et al., 1980). In these cells, too, monensin appears to affect dictysomes and dictysome-derived vesicles specifically. Preliminary ultrastructural observation of Griffithsia rhizoids, that have been recovering for 24 hrs following monensin treatment, suggests that the enlarged cisternae, produced during monensin treatment may fuse with each other, but do not fuse with the plasmalemma. Instead, they appear to be retained within the cytoplasm. Thus these vesicles appear to be permanently affected by monensin treatment. As yet, we have not investigated the effect of monensin on the ultrastructure of shoot cells.

Thus monensin and nigericin reversibly inhibit localized growth in rhizoids and repair shoot cells of Griffithsia and appear to inhibit localized vesicle exocytosis. While they inhibit elongation, these ionophores do not appear to disrupt the organization of the rhizoid tip. Dictysomes and other organelles remain concentrated at the rhizoid apex; the chloroplast band does not disperse. In mammalian cells, monensin has been shown to interfere with processing of molecules in the Golgi apparatus and to prevent their secretion into the extracellular space (Tartakoff and Vassalli, 1978; Tartakoff, 1983). The biochemical basis for the morphological and physiological effects of monensin are not known in all systems. It has been suggested that monensin effects on dictysome structure and function are due to its catalysis of an exchange of Na^+/H^+ across membranes of dictyosomes. This, in turn, may lead to an increase in both intracisternal pH and intracisternal osmotic concentration (Tartakoff, 1983). This may also be the mechanism of action of monensin in Griffithsia. The mechanism of action of nigericin may be similar. In many systems, nigericin catalyzes a K^+/H^+ exchange. However, in Griffithsia effects of nigericin on elongation and cellular morphology appear to be identical to those of monensin. Thus it is likely that both ionophores induce similar changes in ion balance in these cells. In red blood cells both ionophores have been shown to cause a simultaneous influx of Na^+ and efflux of K^+ (Pressman and Fahim, 1982). It seems unlikely that monensin and nigericin effects in Griffithsia are due to changes in internal K^+ levels since their effects on elongation and cytoplasmic organization appear to be very different from those caused by lowered $[K^+]$ or by valinomycin, While both monensin and nigericin might cause similar changes in the ionic and pH relations of dictyosome cisternae, their effects might also be due to the fact that they both cause an increase in cytoplasmic $[Na^+]$. Pressman and Fahim (1982) have suggested that changes in cytoplasmic $[Na^+]$ might lead to changes in free Ca^{+2}, which in turn might mediate changes in organelle distribution and exocytosis. The role of Ca^{+2} in rhizoid and repair shoot cell growth is under investigation.

Thus Griffithsia rhizoids and repair shoot cells are a good model for studying cytoplasmic localization and localized growth. We have found that transcellular ion currents per se do not appear to regulate growth in these cells. K^+, as a osmotic solute, is necessary for continued elongation. Nigericin and monensin have dramatic and similar effects on growth and morphogenesis in all cells of Griffithsia. While the mechanism of action of these ionophores is not known at this time, studies in Griffithsia should help in understanding their effects on plant cells.

Acknowledgements: This work was supported by N.S.F. Visiting Professorships for Women Grant, VPW 8310346, and N.S.F. Grant, DCB 8541028. The author wishes to thank Dr. William Lucas, University of California at Davis in whose laboratory some of this work was begun.

Literature Cited

Bisson, M.A., G.O. Kirst (1979) Aust.J.Plant Physiol. 6, 523-537.

Findlay, G.P., A.B. Hope, E.J. Williams (1969) Aust.J.Biol.Sci. 22, 1163-1182.

Gooday, G.W. (1983) In: Fungal Differentiation: A Contemporary Synthesis. Ed. J.E. Smith, Marcel Dekker, N.Y. pp. 315-356.

Ledger,P.W., N. Uchida, M.L. Tanzer (1980) J. Cell Biol. 87,663 671.

Mollenhauer, H.H., D.J. Morré, R. Droleskey (1983) Protoplasma 114, 119-124.

Nuccitelli, R. (1978) Dev. Biol. 62, 13-33.

Nuccitelli, R. (1983) Modern Cell Biol. 2, 451-481.

Nuccitelli, R., ed. (1986) Ionic Currents in Development, Alan R. Liss, Inc., N.Y. 375 pp.

Pressman, B.C., M. Fahim (1982) Ann.Rev.Pharmacol.Toxicol.22,465-90.

Robinson, D. (1981) Eur. J. Cell Biol. 23, 267-272.

Sievers, A., E. Schnepf (1981) In: Cytomorphogenesis in Plants. Ed. O. Kiermayer, Springer-Verlag, Wien-N.Y. pp. 265-299.

Tartakoff, A.M. (1983) Cell 32, 1026-1028.

Tartakoff A.M., P. Vassalli (1978) J. Cell Biol. 79, 694-707

Waaland, S.D. (1986) In: Plant Browth Substances 1985. Ed M. Bopp. Springer-Verlag, Wien-N.Y. pp. 257-262.

Waaland, S.D., R. Cleland (1974) Protoplasma 79, 185-196.

Waaland, S.D., W.J. Lucas (1984) Protoplasma 123, 184-191.

Waaland, S.D., W. Nehlsen, J.R. Waaland, (1977) Plant & Cell Physiol. 18, 603-612.

Watson, B.A., S.D. Waaland (1983) Plant Physiol. 71, 327-332.

Watson,B.A., S.D.Waaland (1986) Plant & Cell Physiol. 27,1043-1050.

MOLECULAR SIGNALS DURING SEXUAL INDUCTION OF *VOLVOX CARTERI* F. *NAGARIENSIS*

Rainer Gilles, Dirk Balshüsemann, and Lothar Jaenicke
Institut für Biochemie, Universität Köln
An der Bottmühle 2, D 5000 Köln 1, F.R.G.

Sexual reproduction is often induced by environmental stress. Thus, in many unicellular green algae nitrogen starvation triggers the production of thick-walled zygotes which endure long adverse periods. In the multicellular green alga *Volvox carteri*, however, sexual reproduction occurs only under optimal growth conditions and is initiated by a milieu-independent, intrinsic mechanism (Jaenicke and Gilles, 1985). One reason for this divergence might be that - in contrast to other algae in which vegetative cells are directly converted into gametes - *Volvox* spheroids have to undergo transformation into new sexual organisms bearing eggs or sperm before mating can occur (Starr, 1969). This is too long to afford survival in a quickly deteriorating environment. It may be a better strategy to reproduce sexually in time and to store the dormant zygotes before resources become limiting. For this purpose *Volvox* has developed an elaborate sexual induction system consisting of an intrinsic triggering mechanism, a chemical intra-species messenger, and a signal cascade within the spheroids.

The primary event

Volvox synchronizes its sexual reproduction by a pheromone which is released into the medium by sexual males (Starr and Jaenicke, 1974). While this sex inducer is responsible for mass production of gametes, it does not account for the appearance of the first male. In asexually growing cultures of male strains induction of sexuality regularly occurs after the 4th or 5th generation cycle. When this happens, always at least one sexual male was formed in the preceding generation. Due to the high sensitivity of the biological system, it releases sufficient pheromone to turn sexual the whole next generation. The occurrence of the first male is a statistical event with a frequency of 10^{-4} (Weisshaar et al., 1984).

Algal Development (Molecular and Cellular Aspects)
Ed. W. Wiessner, D.G. Robinson, and R.C. Starr

The first male seems to be a mutant in a locus affecting sexuality. Since unmated sperm die, clones of the spontaneous males cannot be obtained. However there is the following circumstantial evidence:

- Female strain form spontaneous sexuals (sex^c) in the same frequency as male strains. Since they do not excrete sex inducer they have no influence on the population. Clones which are constitutive sexuals can be isolated, and the sex^c-trait behaves like a single gene.
- In the male strain we could show that the sex^c-trait is propagated through the cell lineage during embryogenesis.
- In a mutant of the male strain (pan) which forms gonidia instead of sperm packets, the occurrence of spontaneous males has the same frequency of 10^{-4}. Clones of the double mutant (sex^c, pan) are stable for many generations.

The primary event in sexual induction under culture conditions thus occurs as a high frequency mutation. The mechanism which selectively enhances the mutability of the sex^c locus may be a DNA rearrangement as in the yeast mating type switch. Although we can not prove its relevance in the natural habitat, a mechanism initiating sexual reproduction as soon as the population passes over a reasonable density would make sense.

The pheromone

The existence of a sex-inducing substance in the medium of sexual cultures of *Volvox carteri* was shown by R.C. Starr (Starr, 1969). For its bioassay a dilution series of pheromone-containing medium is added to vegetative spheroids. The algae react extremely sensitive to the sex inducer; the limiting concentration is in the range of 10^{-16} M or 10^{-12} g/l (Gilles et al., 1984). Crude media will give 30 to 100 % sexual induction at dilutions of 10^{-8}, i.e. they contain 20-60 µg/l; highly purified pheromone preparations may have titers in the range of 10^{-11}. The very small amount of pheromone produced makes the biochemical characterization of the sex inducer difficult. Large cultures can not be obtained, because, as shown above, the sex^c mutation prevents asexual growth quickly.

The pheromone is a 28 kDa glycoprotein. Due to its high isoelectric point (pI = 10.5) it is easily separated from most *Volvox* macromolecules in the medium which are generally strongly acidic. The purified sex inducer forms two bands on SDS gels which show after renaturation a similar biological activity (Fig. 1). We therefore call them "α- and β-inducer". They have so closely similar properties that classical separation methods failed. By reversed phase HPLC, however, the isolation of the two isoinducers succeeded (Balshüsemann, 1986). Although β-inducer is 1500 Da smaller than α-inducer, it has almost the same amino acid composition as α-inducer (Tab.1).

Table 1: Amino acid composition of α- and β-inducer in mol/mol inducer

amino acid	α-inducer	β-inducer	amino acid	α-inducer	β-inducer
Hyp	-	-	Met	2	3
Asx	19	19	Ile	8	8
Thr	13	14	Leu	16	16
Ser	20	21	Tyr	4	4
Glx	18	18	Phe	11	11
Pro	10	10	Lys	8	8
Gly	13	14	His	1	1
Ala	17	16	Arg	10	10
Cys	5	5			
Val	12	12	$GlcNH_2$	10	10
			$GalNH_2$	-	-

For most amino acids the divergences are within the analytical error, only for methionine the difference of one equivalent is significant. This is further corroborated by cyanogen bromide cleavage which yields at least one more fragment for β-inducer than for α-inducer. In contrast to other glycoproteins in Volvox, the inducer contains no hydroxyproline. Limited proteolysis by V8 protease results in a cleavage pattern which retains the mass difference of the isoinducers in all fragments with the exception of a common product at 13 kDa (Fig.1). This fragmentation indicates a strong sequence homology of α- and β-inducer. After cleavage of the carbohydrate bonds with trifluoromethanesulfonic acid (Edge et al., 1981), a main product at 21 kDa and some smaller fragments presumably resulting from splits of the polypeptide chain are obtained (Fig.1). The common 21 kDa core protein should contain practically no carbohydrate, since the calculated molecular mass based on the amino acid analysis is 20 kDa. We conclude that the major difference between the two molecules is in the carbohydrate portion. Deglycosylation by hydrogen fluoride/pyridine (Coudron, 1980) for 1 hr at 0°C yields polypeptides at 24 and 22,5 kDa for α- and β-inducer, respectively. This mild treatment removes primarily peripheral

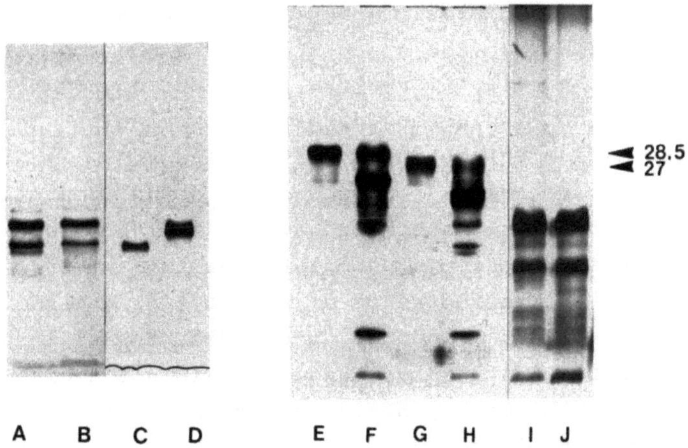

Fig 1: SDS-Polyacrylamide gel electrophoresis of the pheromone

a+b: α- and β-inducer before HPLC separation
c+d: HPLC-separated β-inducer and α-inducer
e+f: α-inducer before and after V8-protease cleavage
g+h: β-inducer before and after V8-protease cleavage
i+j: α-inducer and β-inducer after trifluoromethanesulfonic acid treatment

a-d: 12 % gel; e-j: 15 % gel. All bands were visualized by silver stain.

carbohydrates from the polysaccharide chain, not N- or O-glycosidically bound sugars (Glassman et al., 1978). Since the mass difference of the isoinducers is retained, they may possess a different carbohydrate core region. The chemically deglycosylated pheromone has no sex inducing activity.

These preliminary data suggest that Volvox has two very similar inducer genes possibly derived by a gene duplication, but at least one amino acid exchange has occurred. If by this mutation one glycosylation site in the β-inducer is lost, the difference in the molecular mass could be explained. Proof for this assumption will come from the ongoing sequence analysis.

The effect of the sex inducer

The pheromone alters the timing and differentiation signals of the gonidia during embryogenesis (Jaenicke and Gilles, 1982). Instead of 16 gonidial stem cells, set apart at the 5th cleavage by unequal cell division of distinct blastomeres in the vegetative cycle, after sexual induction 32 egg or 256 sperm stem cells are committed at later stages and other blastomeres are involved in unequal cleavage. Differentiation is not influenced directly, because the inducer must be added to the medium before onset of the cleavages; is not needed, however, during embryogenesis for successful sexual induction (Gilles et al., 1984). The spheroids have to be incubated with the inducer for at least 8 hours for complete reactivity. If during the pre-cleavage phase the pheromone is diluted out or antibodies, raised in rabbits against α-inducer, are added, sexual induction is prevented. This shows that, sexual induction is not decided before the gonidia begin with embryogenesis; after onset of the cleavages they are finally determined.

Our explanation is that the sexual inducer turns on a slow reaction at first reversible upon removal of the pheromone; and that one or several intraspheroidal second messages should exist. Up to now the search for biochemical changes during the first hours of induction gave the following results:

- Immediately after the addition of sex inducer the protein phosphorylation pattern from ^{32}P-phosphate is changed. The phosphorylation of the major phosphoprotein, named pp 290, is drastically down-regulated. After 2 hours two new, induction-specific phosphoproteins, pp 240 and pp 120, are labeled. All three phosphoproteins are secreted into the extracellular matrix (Gilles et al., 1983).
- During sexual induction the level of cAMP increases six-fold in the cells and two-fold in the extracellular matrix. Thus, this cyclic nucleotide may be a candidate for an intraspheroidal differentiation signal (Gilles et al., 1985).
- The pheromone elicits a change in the synthesis of extracellular sulfated glycoproteins (Wenzl and Sumper, 1986). Since these glycoproteins are exclusively synthesized by somatic cells it is not likely that they are involved in sexual induction. A function in transport of the sex inducer to the gonidia has been discussed (Wenzl and Sumper, 1986).

All pheromone-specific reactions are detected in the extracellular matrix. So far no intracellular or membrane protein seems to be affected by the sex inducer. On the other hand there must be some intracellular reaction(s) to explain the altered differentiation of the gonidia. The modifications in the extracellular matrix are obviously not the final response but may play an essential role in the intraspheroidal signal transducing and amplification system.

The role of extracellular matrix

Volvox spheroids contain a highly structured extracellular matrix. There have been some confusing definitions of extracellular matrix in the past, thus it would be helpful to accept the recently proposed system of nomenclature (Kirk et al., 1986) which is also followed in Fig. 2.

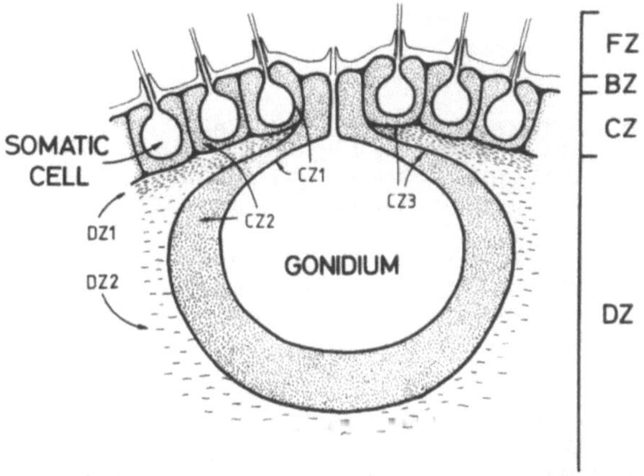

Fig. 2: Schematic drawing of a portion of a Volvox carteri f. nagariensis spheroid showing the extracellular matrix which is divided into four main zones defined as follows: FZ : flagellar zone; BZ : boundary zone, CZ 1-3 : cellular zones; DZ 1-2 : deep zones. The cells themselves are not drawn. Note that the gonidia have a flagellar channel but no flagella.

If the extracellular matrix is essential for sexual induction, isolated gonidia should not be inducible, and evidence for this we had first reported. Reexamination showed, however, that induction of isolated gonidia is more complicated and depends much on the experimental conditions (Wenzl and Sumper, 1986; Gilles, unpublished). As seem from Table 2 inducibility varies with the kind of cultivation of the isolated gonidia. Wenzl and Sumper have found good inducibility on solid starch media and a poor one in liquid culture. They conclude that efficient aeration is essential for successful induction of isolated gonidia, and that the extracellular matrix has no triggering function in sexual induction. In contrast, we find that aeration does not enhance but blocks sexual induction. What really affects inducibility is the amount of medium in which the free gonidia are suspended (Tab. 2.). We therefore postulate that an extracellular substance required for induction is secreted by the cells which should not be diluted out by a large culture volume.

Indeed, drastically reduced induction occurs, if we change the medium of 1 ml cultures several times during the

Table 2: Sexual induction of isolated gonidia under different culture conditions:

kind of cultivation	sexual induction
free gonidia on starch gel	50 %
free gonidia in 0,2 ml medium	40 %
free gonidia in 1 ml medium	20 %
free gonidia in 10 ml medium	5 %
gonidia + somatic cell sheets on starch gel	80 %
gonidia + somatic cell sheets in 0,2 ml medium	80 %
gonidia + somatic cell sheets in 1 ml medium	50 %
gonidia + somatic cell sheets in 10 ml medium	40 %
gonidia + somatic cell sheets in 1 ml medium with aeration	0 %
gonidia + somatic cell sheets in 10 ml medium with aeration	0 %
gonidia + somatic cell sheets in 1 ml medium (with medium change every 2 hr)	5 %

Freshly released spheroids were ruptured by forcing them through a drawn pasteur pipette. Gonidia were separated by filtration through (i) a 50 µm nylon net retaining somatic cell sheets + gonidia, (ii) a 10 µm nylon net retaining free gonidia. Aeration was provided by bubbling with ordinary air.

pre-cleavage phase. The released, yet unknown, substance is also secreted by the somatic cells, since cultivation of mixtures of gonidia and somatic cells yields higher sexual induction levels than that of gonidia alone.

The importance of extracellular matrix for sexual induction cannot be ruled out by isolation of gonidia for two reasons: (i) Isolated gonidia are never free of extracellular material. After physical or enzymatic liberation they are still surrounded by an extracellular structure, the cellular zone 1 (Fig. 2.), which is impossible to remove without killing the gonidia. (ii) Gonidia produce extracellular substances during the experiment. We can demonstrate the release into the extracellular space for two substances, low molecular cAMP and high molecular pp 290, but probably also other substances are secreted. The strongly reduced inducibility of free gonidia indicates that the sexual induction system is severely disturbed. Since the viability of the gonidia is not affected by their isolation, an induction-specific reaction must be affected. This has to be an extracellular event, presumably mediated by

somatic cells. This prooves that the extracellular matrix does play a role in sexual induction of Volvox.
We have calculated from binding studies that the pheromone is concentrated into the extracellular matrix of the spheroids by a factor of 300 (Gilles, et al., 1984). Therefore it is plausible that isolated gonidia need a 100 fold excess of sex inducer for their response (Wenzl and Sumper, 1986). The biochemical basis for this concentrating mechanism may be a specific binding site for the pheromone in the extracellular matrix or an ionic interaction of the basic sexual inducer with acidic extracellular glycoproteins.

Conclusion

The sexual induction system of Volvox carteri is far from being understood on a biochemical level. No receptor for the pheromone, no response of intracellular reactions to the sex inducer, but many modifications of the extracellular matrix have been found, further studies will show whether phosphorylation of extracellular proteins and cAMP are primary molecular signals in sexual induction or only indirectly involved.

References

Balshüsemann, D. (1986) Diplomarbeit, Universität Köln.
Coudron, C.; Ellis, K.; Philipson, L. and Schwartz, N.B. (1980) Biochem. Biophys. Res. Comm. 92, 618-623.
Edge, A.S.B.; Faltynek, C.R.; Hof, L.; Reichert, L.E. and Weber, P. (1981) Anal. Biochem. 118, 131-137.
Gilles, R.; Gilles, C. and Jaenicke, L. (1983) Naturwissenschaften 70, 571-572.
Gilles, R.; Gilles, C. and Jaenicke, L. (1984) Z. Naturforsch. 39, 584-592.
Glassman, J.N.S.; Todd, C.W. and Shively, J.E. (1978) Biochem. Biophys. Res. Comm. 85, 209-216.
Jaenicke, L. and Gilles, R. (1982) in: Biochemistry of Differentiation and Morphogenesis (Jaenicke, L.; ed.) Springer-Verlag, Berlin, Heidelberg, pp 288-294.
Jaenicke, L. and Gilles, R. (1985) Differentiation 29, 199-206.
Kirk, D.L.; Birchem, R. and King, N. (1986) J. Cell Sci. 80, 207-231.
Starr, R.C. (1969) Arch. Protistenk. 111, 204-222.
Starr, R.C. and Jaenicke, L. (1974) Proc. Natl. Acad. Sci. USA 71, 1050-1054.
Weisshaar, B.; Gilles, R.; Moka, R. and Jaenicke, L. (1984) Z. Naturforsch. 39, 1159-1162.
Wenzl, S. and Sumper, M. (1986) Develop. Biol. 115, 119-128.

PHEROMONE-INDUCIBLE GLYCOPROTEINS OF THE EXTRACELLULAR MATRIX OF VOLVOX AND THEIR POSSIBLE ROLE IN SEXUAL INDUCTION

S. Wenzl and M. Sumper
Lehrstuhl für Biochemie I
Universität Regensburg
Universitätsstr. 31
8400 Regensburg

INTRODUCTION

Members of the genus Volvox are simple, multicellular, eukaryotic organisms and each individual is composed of only two types of cells: Somatic and reproductive. Asexual and sexual development of Volvox carteri have been repeatedly described in detail (Starr 1969, 1970). Gamete formation is inducible by the action of a pheromone. In Volvox carteri the sexual pheromone was characterized as a 30 000 - Da glycoprotein (Starr et al. 1974; Gilles et al. 1981), which is synthesized and excreted by sperm cells. The biochemical mechanism of the pheromone's action is almost completely unknown. The ultimate target of the pheromone's action obviously is the gonidium, which responds with production of sexual rather than asexual offspring. However, all biochemical responses to the pheromone detected so far turned out to originate from somatic cells. In this paper we describe pheromone-induced synthesis of sulphated glycoproteins by somatic cells. In addition we discuss possible functions of these extracellular matrix components.

MATERIALS AND METHODS

Culture conditions

Volvox carteri f. nagariensis, strain HK10 (female), was obtained from the Culture Collection of Algae at the University of Texas (R.C. Starr). Synchronous cultures were grown in Volvox medium (Provasoli et al. 1959) at 28°C with a light - dark change of 16 hr light and 8 hr dark (Starr et al. 1974).
All experimental procedures (Pulse-chase labelling, SDS-extraction, SDS-PAGE, fluorography, preparation of Volvox "ghosts",

Algal Development (Molecular and Cellular Aspects)
Ed. W. Wiessner, D.G. Robinson, and R.C. Starr

isolation of gonidia and somatic cell sheets and purification of the sexual inducer) were previously described (Wenzl et al. 1986).

RESULTS AND DISCUSSION

Pheromone induced synthesis of sulphated glycoproteins

A number of extracellular glycoproteins in Volvox are highly sulphated. Thus, pulse-labelling with radioactive sulphate selectively labels these extracellular glycoproteins. Such a pulse labelling experiment performed with Volvox colonies growing either asexually or sexually induced is shown in Fig. 1A. The composition of sulphated glycoproteins becomes quite different after sexual induction: The components F-SG, SSG 110 and SG 70 appear to be marker molecules of sexual development. These sulphated glycoproteins were previously characterized in some detail (Wenzl et al. 1982, 1986).

The change of extracellular matrix composition following the application of sexual pheromone was kinetically analyzed by pulse labelling experiments. For that purpose pheromone was added to a synchronous Volvox population about 10 hr before the onset of embryogenesis (=period of optimal sensitivity for the pheromone). Thereafter aliquots were taken at the times indicated (Fig. 1B) and pulse labelled with radioactive sulphate for only 10 min. The extracts of the labelled Volvox spheroids were analyzed by SDS-PAGE. As early as 10 min after the application of pheromone synthesis of SG 70 becomes detectable. Subsequently, the rate of its synthesis increases, reaches a maximum level about 40 min later. 150 min after the application of pheromone SG 70 synthesis is no longer detectable.

Thus, response to the pheromone lasts only 150 min during the 48 hr life cycle of Volvox. Synthesis of F-SG becomes detectable 30 min after application of the pheromone and persists till early embryogenesis (Wenzl et al. 1982). Synthesis of another sulphated glycoprotein. SSG 140 which is characteristic for asexual development, is turned off following the application of pheromone.

SSG 140 appears to be replaced by a different species, denoted SSG 110.

These drastic changes in the synthetic pattern of extracellular matrix components suggest a functional role for these extracellular molecules in the process of sexual induction.

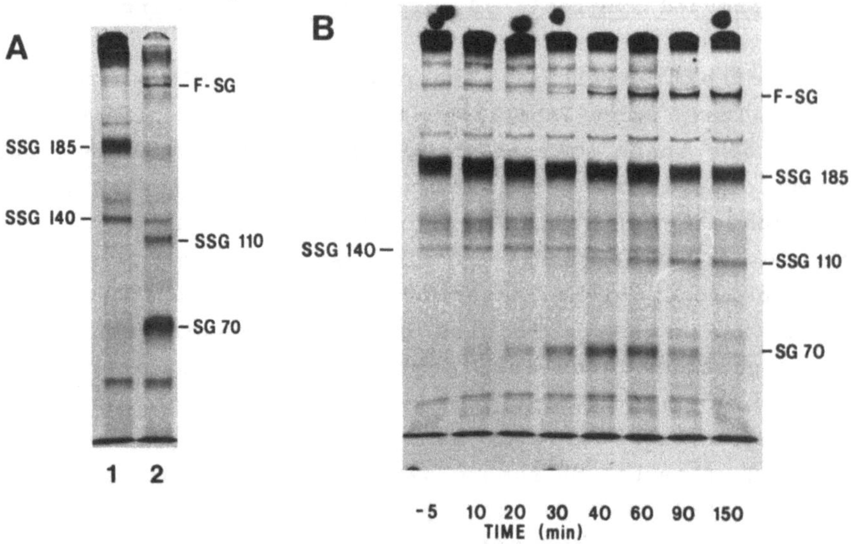

Fig. 1: Pheromone-induced synthesis of sulphated glycoproteins. Fluorograms of SDS-PAGE (6%) loaded with extracts from Volvox-spheroids pulse labelled with radioactive sulphate: A) A synchronously growing Volvox population was divided 10 hr before the onset of embryogenesis. One half served as control (1) the other half (2) was incubated with pheromone for 30 min and subsequently pulse labelled for 30 min. B) Kinetic study of synthesis of pheromone-induced sulphated glycoproteins. After application of pheromone (zero time), aliquots of the Volvox population were pulse labelled for 10 min at the times indicated.

<u>Synthesis of sex-specific sulphated glycoproteins correlate with sexual development</u>

The sexual pheromone of Volvox carteri is one of the most potent biological effector molecules known: It exhibits full effectiveness at 6 x 10^{-17}M (Starr 1970, Gilles et al. 1984). If synthesis of SG 70, SSG 110 and F-SG is indeed one of the events

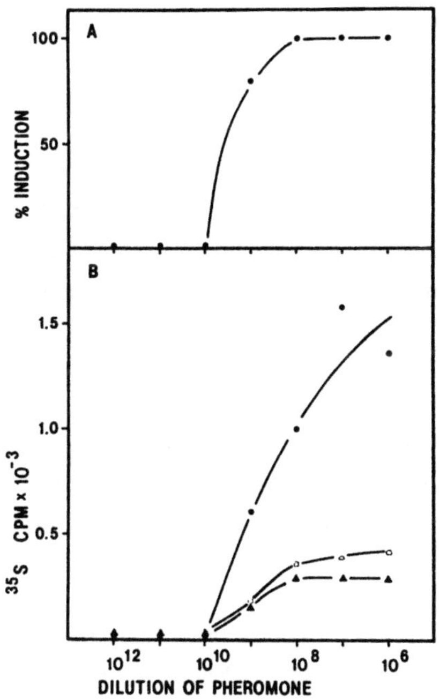

Fig. 2: Dose dependency of sexual induction (A) and pheromone-induced synthesis of sulphated glycoproteins (B). A synchronously growing Volvox population 10 hr before the onset of embryogenesis was treated with pheromone at the dilutions indicated. 30 min later, one half was pulse labelled with radioactive sulphate for 30 min. Extracts were applied to SDS-PAGE. After fluorography F-SG, SSG 110 and SG 70 bands were evaluated quantitatively by liquid scintillation counting (B) • SG 70; ○ SSG 110; ▲ F-SG. The other half of spheroid suspensions were further incubated and the percentage of sexual induction was determined after the end of embryogenesis (A).

causing the reproductive cell to commence sexual development, then synthesis of these sulphated glycoproteins should show the same exquisite sensitivity of response to the pheromone as the overall process of sexual induction. To check this point, a serial dilution experiment was performed. A 10^8-fold dilution of our pheromone stock solution was fully active in the biological assay (i.e., 100 % of a Volvox population was triggered to sexual development). At 10^9-fold dilution, the inducing capacity dropped to less than 80 % and further dilutions remained inactive (Fig. 2A). Therefore we performed $^{35}SO_4{}^{2-}$-labelling experiments in the presence of pheromone concentrations covering the range around the treshold value. The quantitative evaluation of the corresponding labelling patterns with respect to synthesis of the sex-specific glycoproteins is shown in Fig. 2B.

The biosynthetic response to the pheromone is indeed as sensitive as the process of sexual induction. This result supports involvement of these macromolecules in the process of sexual induction.

Cellular localization of sex-specific glycoproteins

For any discussion of the possible functions of the pheromone-inducible glycoproteins it is important to know their localization and site of synthesis. Volvox spheroids easily dissociate into somatic cell sheets and free gonidia upon mild mechanical stress, as exerted by forcing them through a drawn pipette. Purified somatic cell sheets and suspensions of pure gonidia were incubated with the sexual pheromone and 30 min later pulse labelled with $^{35}SO_4^{2-}$. Synthesis of all inducible sulphated glycoproteins took place exclusively within the somatic cell sheets (Fig.3A, lane 2). Pulse labelling of the gonidial suspension did not reveal any synthesis of these components, although a 20-fold amount of gonidia relative to the reference (intact colonies) was applied to SDS-PAGE (Fig.3A, lane 3).

This experimental result is highly unexpected, since the ultimate target of the pheromone's message is the gonidium rather than the somatic cell. Possibly, it is this synthetic activity of the somatic cells which allows a Volvox gonidium to respond to the pheromone with the extreme sensitivity.

Free gonidia can be triggered to sexual development

The observations described so far obviously raize an important question: Is the gonidium able by itself to respond to the pheromone or is the cooperation of somatic cells an absolute requirement for successful induction? It has repeatedly been stated in the literature that free gonidia do not respond to the pheromone, without, however, any presentation of experimental conditions. We therefore reinvestigated the pheromone's action on gonidia in isolation under a number of different experimental conditions. Nearly 100 % of the gonidial population responded to the pheromone and developed to sexual individuals when cultivated on solid media (preferably starch or agarose gels with a low content of sulphated polysaccharides, which might bind the positively charged pheromone). A lower percentage of the gonidial population

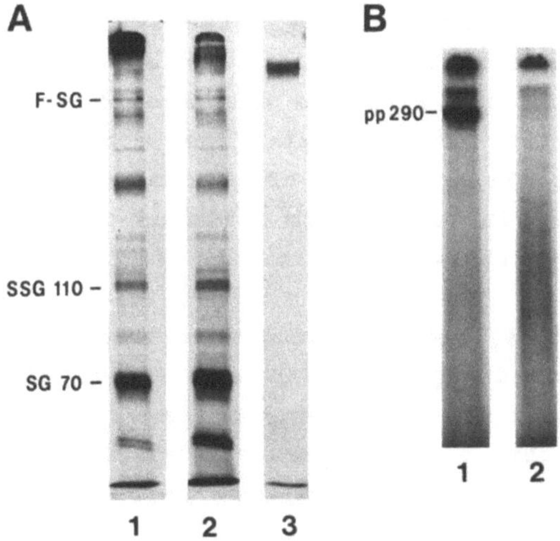

Fig. 3: A) Pheromone-induced synthesis of sulphated glycoproteins by somatic cells. Somatic cell sheets and gonidia were isolated after release from mother colonies. 30 min after the addition of pheromone the samples were pulse labelled with radioactive sulphate (30 min). The extracts were submitted to SDS-PAGE. The corresponding fluorography shows the extracts obtained from: (1) intact spheroids (2) somatic cell sheets, and (3) isolated gonidia. B) Synthesis of pp 290. Gonidia were isolated and after 5 hr incubation under standard culture conditions pulse labelled (30 min) with radioactive phosphate. The extracts were analyzed by SDS-PAGE. The corresponding fluorography shows extracts obtained from: (1) intact spheroids and (2) isolated gonidia.

(about 50 - 60 %) responded to the pheromone if cultivation was performed in suspension using culture petri dishes. No or a very low response was observed if cultivation was performed in test tubes filled to two-thirds with medium, standard conditions for propagation of intact Volvox spheroids. Thus, efficient aeration of isolated gonidia appears to be essential for successful induction of isolated gonidia. Preparations of isolated gonidia are completely free of sulphated glycoproteins (labelled with $^{35}SO_4^{2-}$ prior to the preparation of free gonidia) indicating the complete absence of extracellular matrix material in this preparation. Sexual induction of gonidia requires prolonged incubation in the presence of pheromone. It is thinkable that isolated gonidia recover the capacity for extracellular matrix synthesis during this period. To check this point, isolated free gonidia were first incubated in the presence of pheromone for 5 hr and then pulse labelled with $^{35}SO_4^{2-}$. Again, the gonidia completely lack the capacity for the synthesis of pheromone-inducible sulphated glycoproteins. The incorporation patterns obtained were essentially the same as those shown in Fig. 3A.

Gilles et al. (1983) proposed another extracellular glycoprotein (the phosphorylated glycoprotein pp 290) to be involved in the mechanism of sexual induction. By the same type of experiment, it is demonstrated that this glycoprotein too is exclusively synthesized by somatic cells and completely absent in isolated free gonidia (Fig. 3B, lane 2).

The conclusion to be drawn from these experiments is that all the pheromone-induced changes within the extracellular matrix do not represent an absolute requirement to trigger the gonidium for sexual development. However, there is an experimental observation giving an idea of the function of these molecules. Isolated gonidia respond with a reduced sensitivity to the pheromone. The threshold concentration for the pheromone is two orders of magnitude higher than that for intact Volvox spheroids. Therefore all the pheromone-induced biosynthetic reactions found within the somatic cell sheet could be part of an amplification system accummulating the pheromone and raising the actual pheromone concentration up to 100-fold at the surface of the gonidium. Indeed Gilles et al. (1984) have found a binding capacity for the pheromone in the extracellular matrix. Such a role of the extracellular matrix in the process of sexual induction is quite different from that proposed by a model of Gilles et al. (1984, 1985). These authors assume that a cascade-like signal transforming system exists in the extracellular matrix, which ultimately lowers <u>extracellular</u> levels of cyclic nucleotides. Low cyclic nucleotide levels should then trigger the gonidium. This model was strongly supported by a report that a cyclic phosphodiesterase (c-PDE) added to Volvox spheroids suspension acts as the sexual pheromone. However in our hands, this induction by c-PDE was never reproducible and it now appears that the c-PDE used in the original experiment was contaminated by Volvox pheromone (Gilles, personal communication), an event which happens regularily in laboratories working with high pheromone concentrations.

A second argument against the action of a second messenger is the successful sexual induction of isolated free gonidia. This latter observation favours the idea that the pheromone by itself triggers the gonidium for sexual development.

ACKNOLEDGEMENT
This work has been supported by the Deutsche Forschungsgemeinschaft (SFB 43).

REFERENCES

Gilles, R., C. Bittner, L. Jaenicke (1981) FEBS Lett. 124, 57-61.
Gilles, R., C. Gilles, L. Jaenicke (1983) Naturwissenschaften 70, 571-572.
Gilles, R., C. Gilles, L. Jaenicke (1984) Z. Naturforsch C. 39, 584-592.
Gilles, R., C. Gilles, L. Jaenicke (1985) FEBS Lett. 184, 309-312.
Provasoli, L., I. J. Pintner (1959) In: The Ecology of Algae, ed. Tyron, C. A., R.T. Hartmann, University of Pittsburgh. pp. 84-96.
Starr, R.C. (1969) Arch. Protistenk. 111, 204-222.
Starr, R.C. (1970) Dev. Biol. Suppl. 4, 59-100.
Starr, R.C., L. Jaenicke, (1974) Proc. Natl. Acad. Sci. USA 71, 1050-1054.
Wenzl, S., M. Sumper (1982) FEBS Lett. 143, 311-315.
Wenzl, S., M. Sumper (1986) Dev. Biol. 115, 119-128.

ENVIRONMENTAL AND PHEROMONAL CONTROL OF SEXUAL REPRODUCTION IN LAMINARIA (PHAEOPHYCEAE)

Ingo Maier
Fakultät für Biologie
Universität Konstanz
D-7750 Konstanz, F.R.G.

Introduction

Laminaria species form extensive beds on marine rocky coasts of mainly northern hemisphere cold-temperate regions. For example, three species are found together at Helgoland (North Sea): *L. digitata*, *L. hyperborea*, and *L. saccharina*, inhabiting the sublitoral region in about 0-4 m below mean low water spring tide. *L. hyperborea* is restricted to the deeper region of the kelp zone (1-4, max. 8 m below M.L.W.S.T. at Helgoland) [Lüning 1985]. In this plant community, the large *Laminaria* sporophytes are the ecologically dominating organisms.

Reproduction in *Laminaria* obligatory involves sexuality. In its life history, the sporophytes alternate with microscopic gametophytes of either male or female sex. Gametogenesis is strictly controlled by environmental factors such as light and temperature. Finally, to secure fertilization, a highly sensitive and very specific communication by a chemical messenger, a pheromone, is realized between gametes in those species (Fig. 1).

Based mainly on experimental observations on *L. digitata*, a szenario on sexual reproduction in *Laminaria* is developed, covering following aspects:

- Gametogenesis and egg release
- Pheromone secretion
- Spermatozoid release: light and scanning electron microscopy
- Pheromone receptor specificity and sensitivity for spermatozoid release
- Physiological studies on spermatozoid release
- Chemotaxis and fertilization
- Interspecific cross-reactions

Algal Development (Molecular and Cellular Aspects)
Ed. W. Wiessner, D.G. Robinson, and R.C. Starr

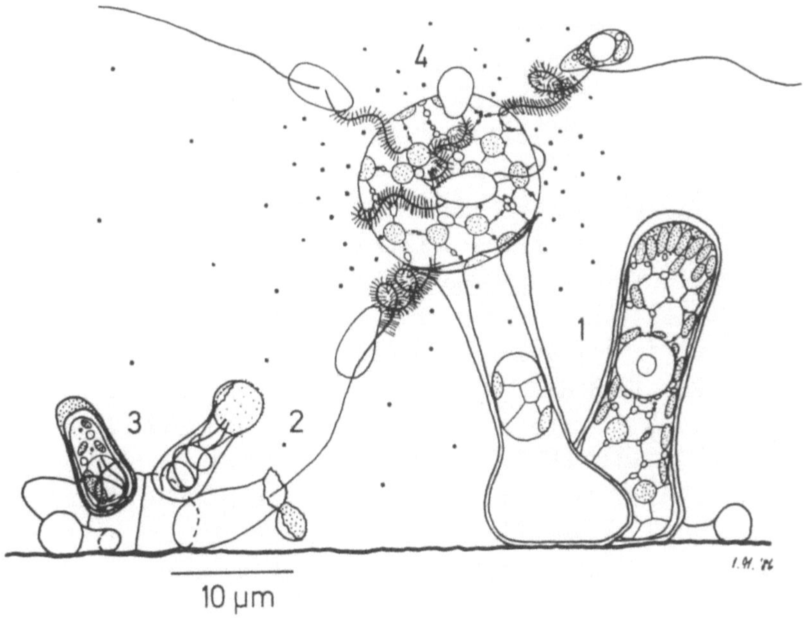

Fig. 1 Pheromonal interaction in sexual reproduction in Laminaria. Eggs released from oogonia (1) secrete a pheromone which causes spermatozoid release (2) from antheridia (3) and, subsequently, gamete attraction (4).

Gametogenesis and egg release

Sporophytes may produce some billions of spores. After settling, spores form a cell wall and germinate immediately. In the one-cell stage, gametophytes may survive the winter months even in complete darkness. In respect of photosynthesis and vegetative growth, laminarian gametophytes can be characterized as extreme-shade plants. Vegetative growth in L. digitata gametophytes is light-saturated at 3 nMol $cm^{-2}s^{-1}$ (continuous white fluorescent light). Gametogenesis, however, requires a relatively high quantum dose of blue light at low temperature. At $5^{o}C$ 50% of the L. digitata female gametophytes become fertile after receiving a quantum dose of 9×10^5 nMol cm^{-2} blue light (400-512 nm). Male gametophytes behave similarly [Lüning 1980].

At Helgoland, the sea is highly turbid until the end of January and light is not sufficient for gametogenesis. When the water becomes calm and clear in February [Lüning & Dring 1979], gametophytes of the three species, L. saccharina, L. hyperborea and L. digitata at 2 m depth mature within 24 days at $5^{o}C$ water temperature [Lüning 1980]. The female gametophytes become fertile in the unicellular stage, whereas the male ones may consist of a few cells. Antheridia mature 1-2 days earlier than oogonia. At least in given depth zones, sexual reproduction in those algae may thus be restricted to few days of the year.

Under culture conditions, Laminaria gametophytes can be grown vegetatively at elevated temperatures and very low photon flux density ($18^{o}C$, 0.3 nMol $cm^{-2}s^{-1}$ white fluorescent light, 18 hrs daylength) infinitely. Synchronized gametogenesis then can be induced by transfer into relatively strong white fluorescent light and low temperature conditions ($10^{o}C$, 3 nMol $cm^{-2}s^{-1}$, 12hrs daylength).

So far known, egg release in all Laminariales occurs during the night, beginning a few minutes after onset of darkness [Lüning 1981, Maier 1986]. It is inhibited by blue light and U.V. irradiation [Lüning 1981]. Egg release can be observed in a microscope under illumination with green or red light. The oogonium wall ruptures at the base of the oogonium cap, a characteristic apical thickening of the cell wall. The egg is then extruded within about 30 seconds.

Pheromone secretion

Immediately after release, eggs begin to secrete volatile, odoriferous chemical messengers, which, after perception by a spermatozoid, cause its release and, subsequently, chemotaxis. Egg secretions may contain a complex bouquet of compounds. In Laminariales, 16 components have been identified so far, one of them being the highly active and specific sexual pheromone [Maier et al., in prep.]. The Laminaria pheromone is an epoxy-buten-cycloheptadiene, named lamoxirene (Fig. 2) [Marner et al. 1984]. Most by-products identified so far are also volatile and hydrophobic C_{11}-hydrocarbons which, like ectocarpene, desmarestene and multifidene, are known as pheromones in other brown algae [Maier & Müller 1986].

Brown algal pheromones are biologically synthesized from unsaturated fatty acids [Boland & Mertes 1985]. The cellular site of synthesis and details of pheromone storage, as well as mechanism, rate and period of secretion are unknown. It can be estimated from pheromone-extraction yields that an average egg produces up to 1×10^{-11} g of lamoxirene.

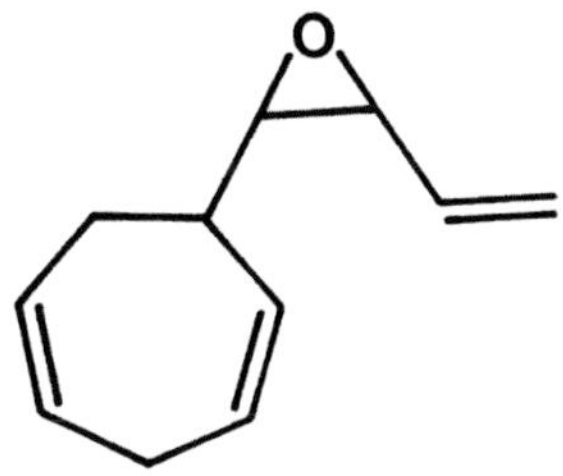

Fig. 2 Structural formula of lamoxirene

Spermatozoid release: light and scanning electron microscopy

A few seconds after pheromone perception, spermatozoids are released from antheridia by a dramatic reaction [Lüning & Müller 1978, Müller et al. 1985]. A male gametophyte of 0.6 mm diameter may produce over 100 000 antheridia. Each antheridium contains one spermatozoid. Antheridium dehiscence takes place within a split second and the process cannot be resolved by the eye. Therefore, it has been investigated by high-speed cinemicrography (Fig. 3, in cooperation with the Institut für Wissenschaftlichen Film, Göttingen, F.R.G.).

In the apical region of the antheridium the cell wall is markedly thickened like in oogonia to form a "cap" [Maier & Müller 1982, Maier 1982]. During dehiscence, the wall in this region is disintegrated to such an extend, that the spermatozoid can rupture it within about 0.3 seconds and then is forced out of the antheridium (Fig. 3), apparently driven by an internal pressure which perhaps is generated by swelling of mucilage [Maier 1982]. The flagella are passively dragged behind until the spermatozoid starts moving at about 1 secound after the begin of release (Fig. 3).

Pheromone receptor specificity and sensitivity for spermatozoid release

The lowest pheromone concentration at which spermatozoid release can be triggered in the most sensitive cells is 1×10^{-11} M [Müller et al. 1985]. At this threshold concentration, release occurs after about 45 s, at 1×10^{-9} M after 11 s.

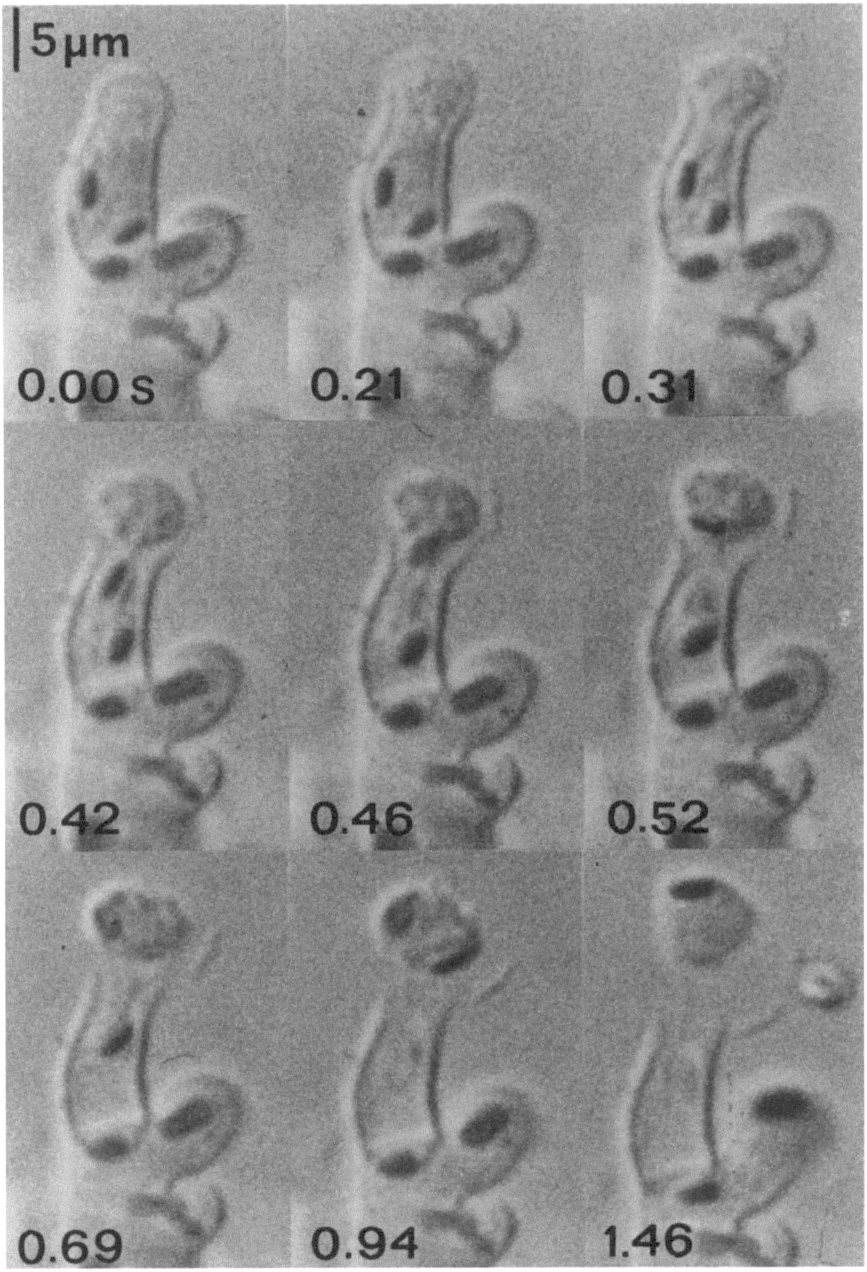

Fig. 3 Successive stages in spermatozoid release in Laminaria digitata. Single frame copies from a high-speed film.

Various synthetic pheromone analogues (kindly provided by W. Boland and L. Jaenicke, Universität Köln, F.R.G.) have been tested for activity. None of them reaches the potency of the natural compound. Preliminary results indicate that any change of the molecule as, for example, hydrogenation of a double bond, drastically reduces activity [Maier et al., in prep.]. Binding to the receptor is most probably achieved by van der Waals' interactions, electrostatic forces between electrons of the pheromone molecule and positive charges at the receptor site [Boland et al. 1981, Maier & Müller 1986].

Physiological studies on spermatozoid release

To the present state of knowledge the cellular processes following pheromone binding are subject to speculation. They must include signal transduction, steady state level determination of some kind of "second messenger" (or perhaps an "integration" over time intervals) and, by reaching a critical level, triggering of a sequence of events which finally leads to breakdown of the antheridium wall and spermatozoid liberation.

When discussing pheromone binding, the location of the postulated pheromone receptor was neglected. In principle, two hypotheses are possible:

1. The receptor is situated in the antheridium wall or in the space between cell wall and cytoplasmatic membrane. In the first case, the receptor is bound to cell wall components, in the latter it may be able to diffuse within the mucilage.

2. The receptor is either integrated in the gamete's cytoplasmatic membrane and is exposed to the extracellular medium or is located inside the cell, either membrane-bound or not.

Sexual pheromones of brown algae are generally low-molecular weight (ca. 150 d) hydrophobic compounds and therefore, like steroid hormones, may easily diffuse through membranes and bind to intracellular receptors.

It has been shown by variation of the ionic composition of the extracellular medium and substitution experiments that the release process depends on certain cations, especially Na^+ and Ca^{2+} or Mg^{2+} in the extracellular medium. Moreover, transient cation influxes following pheromone binding have been supposed [Maier 1982]. Treatment of antheridia with the Ca^{2+}-Ionophor Calcimycin (A 23187, 10 µM, 10 min incubation) results in inhibition of pheromone triggering and slow spontaneous release. Trifluoperazine (5 µM), known to block Ca^{2+}-dependent processes by interaction with Calmodulin, causes practically complete dehiscence of antheridia. The released spermatozoids disintegrate. Saponine treatment (0.5 mg/ml) also causes complete release, but only if Ca^{2+} is present in the extracellular medium. Cyanide (10 mM, 5 min) and 2-dinitrophenol (1 mM, 10 min) block spermatozoid release.

In the whole, these results imply that metabolic and regulatory processes within the sperm cell are involved in the pheromone-triggered release reaction. If one assumes that the pheromone receptor is situated outside the cell (hypothesis 1), an additional signal transmission system has to be postulated, which transmits information from the receptor to the cell by a diffusing "second messenger". That would mean an unnecessary and unreasonable complication and therefore, I suppose the receptor to be located in the cell or at the cell's surface (hypothesis 2). Moreover, this is strengthened from an evolutionary point of view, because the pheromone receptor for sperm release certainly has been derived from a chemotaxis-receptor.

In any case, signal transduction and processing in the sperm cell finally must result in a breakdown of the mechanical integrity of the cell wall by a presently unknown, probably enzymatic process. As yet, I have not been successful in dissolving the antheridium wall by treatment with commercially available enzymes and mixtures of them. From cytochemical studies on 0.5 µm plastic sections it is known that the antheridium cap consists mainly of sulphated polysaccharides, whereas other parts of the cell wall are composed of alginates, some kind of sulphated polysaccharides and perhaps cellulose [Maier 1982].

Chemotaxis and fertilization

Following release, spermatozoids are attracted by eggs. They are guided alond a gradient of the same signal substance which has triggered their release, lamoxirene.

Spermatozoids show a strong chemophobic locomotion behaviour: perception of a decreasing pheromone concentration causes them to reorientate towards the pheromone source. The gametes move by meandering their anterior flagella in helical or, by swimming along a surface, planar waves. The average speed is about 50 µm s^{-1}. Abrupt changes in the direction of movement (angles larger than 90°) are brought about by a sudden 60-100 ms beat of the long posterior flagellum at its base. Spermatozoids seem to be unaffected by pheromone at isotropic conditions, but preliminary results indicate that a negative pheromone gradient increases the beat frequency of the posterior steering flagellum.

An interesting aspect would be to know how many pheromone molecules are necessary to excert a gamete's behaviour response. The problem is not trivial at all and as yet the question cannot be answered satisfactorily. A first approach can be made by quantitative chemotaxis-assays. Such experiments are based in principle on measuring the accumulation of gametes on the surface of solvent droplets containing pheromone at known concentrations, compared to pure solvent [Maier & Müller 1986]. It was found that spermatozoids of *Laminaria digitata* accumulate significantly at droplets with as low as about 1×10^{-12} M lamoxirene at the surface [Maier et al., in prep.]. At this threshold concentration, as a theoretical maximum about 10 molecules hit the surface of a spermatozoid per second!

Gamete recognition in a strict sense does not depend on pheromonal communication, but probably on species-specific recognition structures and complementary receptors on the gametes' surface membranes [Bolwell et al. 1979]. Fusion takes place within a few seconds.

Interspecific cross-reactions

As a sexual pheromone, lamoxirene is common to most members of the brown algal order Laminariales [Müller et al. 1985]. Therefore, interspecific cross-reactions in spermatozoid release and chemotaxis have to be expected in those cases, where different species are living in close vicinity and reproduction periods overlap, as in the European *Laminaria* species. Such interactions can easily be demonstrated in experiments. By-products in egg secretions may also affect gamete orientation in other competitors. For example, *Ectocarpus* male gametes are attracted by *Laminaria* eggs. The significance of such a "chemical competition" in nature is unknown.

References

Boland, W., K. Jacoby, L. Jaenicke, D.G. Müller, E. Fölster (1981) Z. Naturforsch. 36c, 262-271.

Boland, W., K. Mertes (1985) Eur. J. Biochem. 147, 83-91.

Bolwell, G.P., J.A. Callow, L.V. Evans (1979) J. Cell Sci. 36, 19-30.

Lüning, K., D.G. Müller (1978) Z. Pflanzenphysiol. 89, 333-341.

Lüning, K., M.J. Dring (1979) Helgoländer wiss. Meeresunters. 32, 403-424.

Lüning, K. (1980) J. Phycol. 16, 1-15.

Lüning, K. (1981) Br. phycol. J. 16, 379-393.

Lüning, K. (1985) Meeresbotanik. G. Thieme, Stuttgart, pp.44-60.

Maier, I. (1982) Protoplasma 113, 137-143.

Maier, I., D.G. Müller (1982) Phycologia 21, 1-8.

Maier, I. (1986) Sexualität bei Braunalgen aus der Ordnung Laminariales Hartung-Gorre, Konstanz.

Maier, I., D.G. Müller (1986) Biol. Bull. 170, 145-175.

Marner, J.-J., B. Müller, L. Jaenicke (1984) Z. Naturforsch.39c, 689-691.

Müller, D.G., I. Maier, G. Gassmann (1985) Phycologia 24, 475-477.

ACTION OF CELL WALL AUTOLYSINS IN ASEXUAL REPRODUCTION OF FILAMENTOUS GREEN ALGAE: EVIDENCE AND SPECIES SPECIFICITY

U.G. Schlösser
Sammlung von Algenkulturen, Pflanzenphysiologisches Institut der Universität, D-3400 Göttingen, F.R.G.

Simple filamentous green algae do not differentiate into somatic and reproductive parts; all, or nearly all, cells are able to reproduce. They might be suited for investigations on reproduction physiology because, unlike unicellular green algae, vegetative growth is not necessarily connected with a reproduction process. In cultures the vegetative development (cell growth and divisions within the filaments and occasional fragmentation into long filaments) can be strictly separated from reproduction (formation and liberation of reproduction units, whether or not preceded by differentiating nuclear divisions). The following modes of reproduction are possible, according to the species:

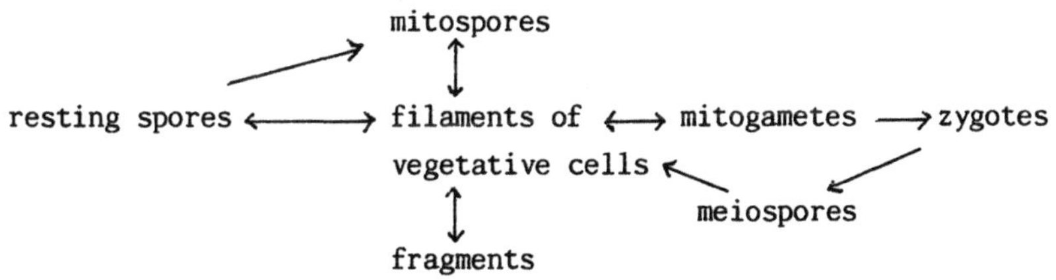

The morphology of these temporary events were described more than 100 years ago. Investigations on their physiological basis, which is usually not known, require synchronization of the reproduction step in larger quantities of material.

This contribution deals with two aspects of asexual reproduction in three genera of filamentous green algae: induction of the reproduction process in cultures, and the participation of special cell wall autolysins.

The conditions used for vegetative growth were similar for the three genera. Filaments were grown in aerated liquid mineral medium by daily light-dark-changes (light:dark= 14:10 h, 2000 lux, compressed air + 2% CO_2). Medium and culture chamber were the same as developed for *Chlorella* (Kuhl and Lorenzen, 1964).

Uronema confervicola SAG 386-2

Uronema has unbranched filaments of indefinite length and differs from the well-known and similar genus *Ulothrix* in two distinctive attributes (Pankow, 1960): a discoidal jelly-like holdfast of the basal cell (no rhizoids formed)

Algal Development (Molecular and Cellular Aspects)
Ed. W. Wiessner, D.G. Robinson, and R.C. Starr

and the irregularly lobed and often perforated chloroplast (no girdle-shaped plates). Reproduction is known only by quadriflagellate spores, one in each cell, which are discharged through a hole in the longitudinal wall. At 20° C and dilution 1:1 with new nutrient solution every week cultures only grow vegetatively and the filaments fragment into longer pieces. The differentiation of spores needs special inducing conditions.
Sporulation can be induced after 3-4 weeks of cultivation by the following changes: transfer from light-dark-change to continuous light in aerated culture for two days, changing the nutrient solution every day, and raising the temperature to 23° C, subsequently transferring to continuous dark in the aerated culture chamber. Almost quantitative sporulation starts 10-12 h later and is finished within 1-2 h.

Evidence of cell wall autolysin activity during sporulation .
The medium after sporulation was made cell-free by centrifugation and tested in a bioassay: 1 ml of a dense culture of heat-fixed (5 min 55° C) filaments containing vegetative cells were incubated at 30° C with 4 ml of the sporulation medium. Within about 2 h 70-95% of the killed protoplasts were liberated (fig.1) by dissolution of the longitudinal walls, but not of the cross walls (fig. 2). The sporulation medium contains a heat-labile lytic factor which we shall term "autolysin" as long as its properties are unknown (Schlösser, 1981). This also induces protoplast discharge in living filaments, but more slowly and less quantitatively. The protoplasts burst under the influence of a detergent (1% Triton X100) within 3-5 min. Regeneration of a new cell wall and further development have not yet been achieved.

Specificity of autolysin action . The autolysin is not developmental stage-specific in its action; during sporulation it dissolves the wall of the sporangium stage and in the bioassay that of the vegatative cell.

The autolysin activity of Uronema confervicola is not confined to this species and not even to the genus Uronema . The following strains of the Sammlung von Algenkulturen Göttingen (SAG) react in the bioassay:

109.80	Chaetophora elegans	410-1	Stigeoclonium pascheri
413-1	Chaetophora incrassata	477-3	Stigeoclonium sp.
413-2	Chaetophora sp.	33.86	Uronema acuminata
111.80	Draparnaldia mutabilis	34.86	Uronema belkae
418-1	Draparnaldia plumosa	386-2	Uronema confervicola
112.80	Fritschiella tuberosa	53.86	Uronema elongatum
8.85	Hormidiospora verrucosa	386-3	Uronema gigas

7.85	Stigeoclonium aestivale	47.85	Uronema trentonense
477-1	Stigeoclon.helveticum	477-23	Uronema sp.
477-2	Stigeoclonium helveticum		

The genera Chaetophora, Draparnaldia, Hormidiospora, and Uronema were induced to liberate protoplasts, whereas in Fritschiella and Stigeoclonium filaments were fragmented. The following strains of the SAG were not effected by the autolysin of Uronema confervicola in the bioassay:

119.80	Cephaleuros virescens	115.80	Microthamnion strictissimum.
110.80	Coleochaete scutata	26.83	Prasiolopsis ramosa
467-2	Dilabifilum arthopyreniae.	466-1	Pseudendoclonium basiliense
49.80	Geminella sp.	467-1	Pseudendoclonium printzii
20.84	Geminella sp.	465-1	Pseudendocloniopsis botryoides.
21.84	Geminella sp.	27.83	Radiofilum transversale
335-1a	Klebsormidium flaccidum	44.84	Schizomeris leibleinii
335-7	Klebsormidium subtil.	118.80	Trentepohlia sp.
445-1	Leptosira obovata	84.81	Trichophilus welkeri
113.80	Leptosira obovata	37.86	Ulothrix tenuissima
348-1	Microspora amoena	38.86	Ulothrix zonata
114.80	Mirothamnion curvatum		

Radiofilum transversale SAG 27.83

was grown vegetatively at 25° C with 5µg/l vitamin B_{12} added to the nutrient solution; the medium was replaced every week. This species has non-polarized filaments with a jelly-like matrix (fig.3) and fragments into shorter filaments, which is the only known mode of reproduction. However in the type species of the genus, R. conjunctivum, Schmidle (1894) had observed flagellated spores in material from nature. The present investigation confirms the formation of spores in cultures of R. transversale.

Sporulation can be induced by nitrogen deficiency and a change of the light-dark-regime carried out in three steps:

1. Transfer into an "inducing medium", containing only the phosphate-buffer, $MgSO_4$ and $CaCl_2$ in concentrations of the nutrient solution, aerated culture for 24 h, light-dark = 14:10 h.
2. 3 x washing in deionized water, transfer of the sediment into Petri dishes of 20 cm ∅ for 24 h in continuous light of 2000 lux, 20° C. During this phase the jelly swells (fig. 4) a feature dependent upon the living condition because the jelly of killed filaments (10 min 55° C) does not swell.
3. Transfer of the Petri dishes into continuous darkness, 20° C. During this phase the jelly dissolves totally within 24-48 h and the proper cell wall

swells after this and during differentiation of the protoplast into a single biflagellated spore (fig. 5,6).

Thus the change of the wall during sporulation may be described as a two step reaction: total dissolution of the jelly and swelling of the sporangium wall which opens in a cup-shaped fashion and dissolves later as well . The liberated naked spores, which at first are long, soon become rounded, settle and after production of a new cell wall grow into a new filament. Addition of KNO_3 in the concentration of the nutrient solution at the beginning of step 3 completely prevents the dissolution of the jelly and the differentiation of spores.

Autolysin activity and its species specificity were evident in the bioassay: cell-free centrifuged sporulation medium (4 ml) dissolved the jelly of living or fixed filaments (1 ml, 5 min 55° C) at 30µ C within 2-6 h (fig. 7-10). The proper cell wall however was not dissolved, thus protoplasts were not discharged. It may be that the autolysin acts on this developmental stage specifically and dissolves only the wall of the sporangium stage. This is e.g. the case in the jelly-producing Chlamydomonas frankii , where a sporangium wall autolysin can dissolve the wall of the sporangium stage and not that of spores or vegetative cells; whereas the jelly outside the proper wall could be dissolved in all developmental stages during the asexual cell cycle (Schlösser, 1976).

The Radiofilum transversale autolysin acts species-specific, it does not dissolve the filament jelly of Radiofilum conjunctivum SAG 9.85, and also not the cell walls of any other genus tested, e.g. Uronema, Geminella , and Chlamydomonas spp. It is heat-labile (inactivated after 1 min at 100° C.)

Geminella sp. SAG 49.80

The non polarized filaments of Geminella are also surrounded by a jelly sheet (fig. 11), differing from Radiofilum by another type of cell wall. The genus reproduces only by fragmentation into shorter filaments.

Fragmentation into single cells can be induced by nitrogen deficiency and cessation of aeration in two steps:

1. Transfer after three weeks in aerated culture at 25° C and dilution with fresh nutrient solution 1:1 every week into the "induction medium" as used in Radiofilum and further 24 h aerated culture at 25° C and light-dark-change 14:10 h.
2. 2 x washing with deionized water and transfer 10 x concentrated in Petri dishes of 20 cm into continous light. During this phase the filaments fragment nearly quantitatively into unicells (fig. 12) within 24 h; however the jelly does get dissolved only partially. Fragmentation is quantitatively

inhibited during this phase after addition of KNO_3 or NH_4Cl (10^{-2}M/1), cycloheximide 20 ug/l), or EDTA (10^{-3}) to the washed filaments.

The cell-free centrifuged medium after fragmentation contains a "fragmentation-autolysin". In the bioassay, killed filaments (from 1 ml of a dense culture, fixed 5 min at 55° C, incubated at 30° C with a 4 ml fragmentation medium) were fragmented into unicells and short fragments within 10-16 h. The heat-labile autolysin does not withstand 5 min at 70° C.

Specificity of the fragmentation autolysin. The autolysin also fragments fixed filaments of four other isolates of *Geminella* from different geographical regions in the bioassay: SAG Nr. 20.80, 105.80, 54.81, 21.84. The autolysin had no effect visible in the microscope on any other genus tested, e.g. *Uronema*, *Radiofilum*, *Chlamydomonas* spp.; it seems to be specific to the genus *Geminella*.

Conclusion

The cell wall autolysins have several potential usages:

- Their further investigation may contribute to understanding reproduction mechanisms. Developmental stage specific autolytic changes of cell walls seems to be characteristic for the discharge of reproduction units in most plants.

- Autolysins might be used as special tools for the analysis of cell wall structure at the molecular level by hydrolysis into fragments of natural composition. Corresponding to the very specific actions of the autolysins of filamentous green algae and *Chlamydomonas* spp. (Schlösser, 1976) one has to expect large diversities in the molecular structure of the jellies and walls.

- Autolysins were used for the isolation of pure cultures. Autolysin induced jelly-free unicells of *Radiofilum* and unicellular fragments of *Geminella* could be purified by washing and streaking on agar dishes.

- The species-specificity of the autolysins can be used as chemotaxonomic markers (Schlösser, 1984). This is of special interest in the green algae where the classification is going to be changed fundamentally (Mattox and Stewart, 1984). One example: the genera *Uronema* and *Ulothrix* were often united in the genus *Ulothrix* (Mattox and Bold, 1962). However according to electron microscope data on the flagellar root system and the cell division process (Sluiman et al., 1980, 1983) both genera will have to be separated again and put into different classes, *Ulothrix* to the Ulvophyceae and *Uronema* to the Chlorophyceae (sensu Mattox and Stewart, 1984). The specificity of the Uronema autolysin supports the separation of the two genera: it acts on Uronema spp. but does not effect typical Ulothrix spp. like U.tenuissima and U. zonata.

References

Kuhl,A., H.Lorenzen (1964) In: Methods in Cell Physiology. Ed.D.M.Prescott, Academic Press, London, New York, p.p. 159-187

Mattox,K.R., H.C.Bold (1962) Phycological Studies III, Univ. Texas Publ. No. 6222

Mattox,K.R., K.D.Stewart (1984) In: Systematics of the Green Algae. Ed. D.E.G. Irvine, D.M.John, Academic Press, London, Orlando, pp. 29-72

Pankow, H. (1960) Arch. Protistenk. 105 , 117-130

Schlösser, U.G. (1976) Ber. Dtsch. Bot. Ges. 89 , 1-56

Schlösser, U.G. (1981) In: Encyclopedia of Plant Physiology N.S., 13B , Plant Carbohydrates II, ed. W. Tanner, F.A. Loewus, Springer, Berlin, Heidelberg, pp. 333-351

Schlösser, U.G. (1984) In: Systematics of the Green Algae. ed. D.E.G. Irvine, D.M.John, Academic Press, London, Orlando, pp. 409-418

Schmidle, W. (1984) Flora 78 , 42-66

Sluiman, H.J., K.R.Roberts, K.D.Stewart, K.R.Mattox (1980) J. Phycol. 16 , 537-545

Sluiman, H.J., K.R.Roberts, K.D.Stewart, K.R.Mattox, G.M.Lockhorst (1982) J. Phycol. 18 , 1-12

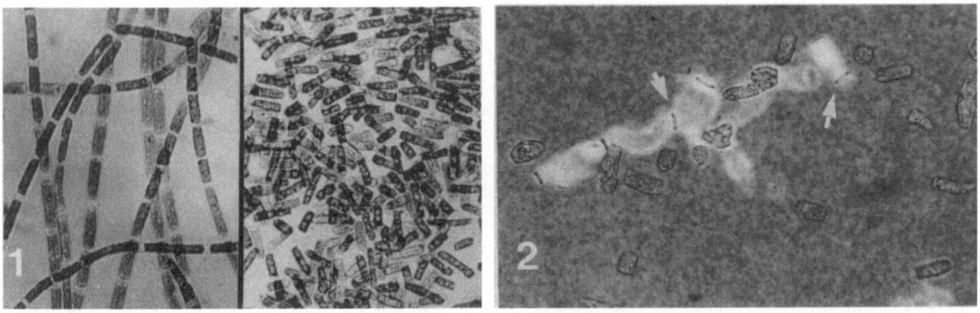

Ulothrix confervicola SAG 386-2
liberation of protoplasts from heat-fixed filaments in the bioassay after addition of cell wall autolysin, contrasted with Indian ink.
fig. 1: filaments and liberated protoplasts after 2 h incubation
fig. 2: liberated protoplasts, dissolving longitudinal walls, cross walls not dissolving (arrows)

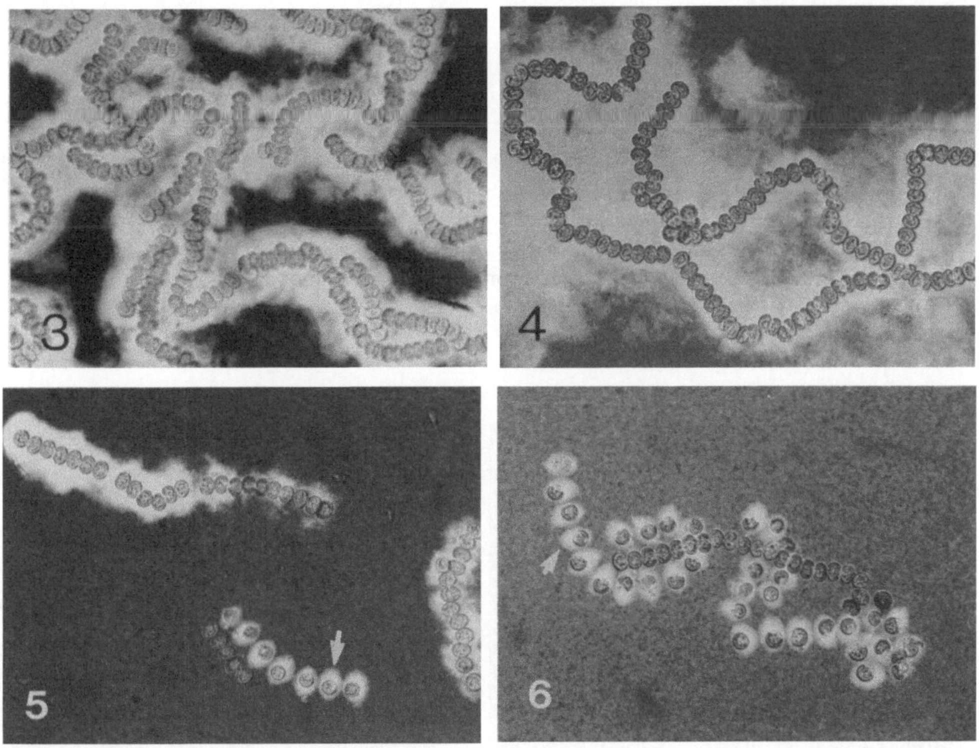

Radiofilum transversale SAG 27.83
change of filament jelly and of cell walls after induction of sporulation, contrasted with Indian ink.
fig. 3: filaments from aerated culture, fig. 4 swelling of the jelly,
fig. 5: dissolution of the jelly, figs. 5,6: jelly-free filaments and swelling sporangia walls (arrows).

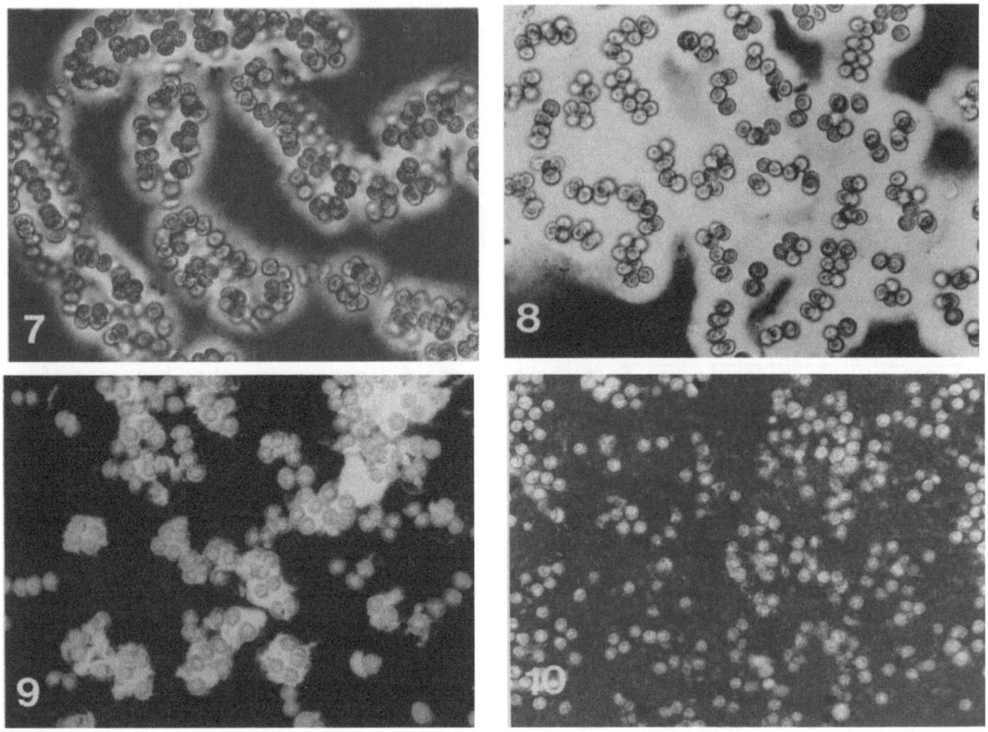

Radiofilum transversale SAG 27.83
figs. 7-10: swelling and dissolution of the jelly of heat-fixed filaments in the bioassay after addition of autolysin, contrasted with Indian ink.

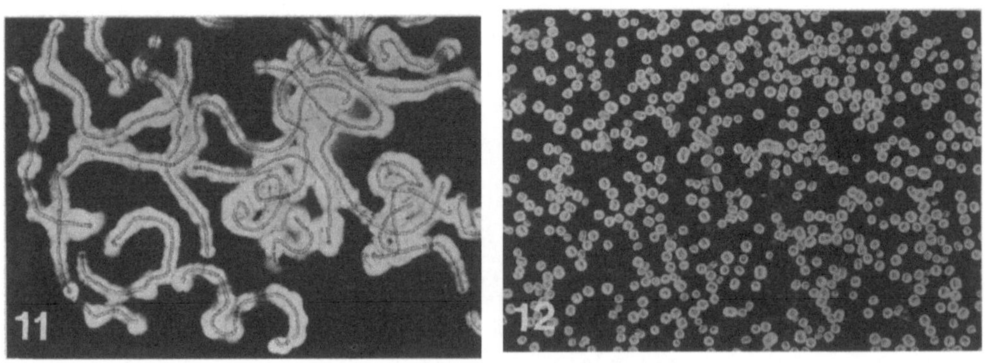

Geminella sp. SAG 49.80
fig. 11: the jelly producing filaments
fig. 12: induced fragmentation into unicells
contrasted with Indian ink

SEXUAL AGGLUTINATION IN CHLAMYDOMONAS EUGAMETOS

A. Musgrave, Dept. of Plant Physiology
University of Amsterdam, 1098 SM Amsterdam

Sexual reproduction in the unicellular green alga Chlamydomonas is proving to be one of the most successful plant systems for studying cell-cell recognition in all its different facets. The two most popular species under investigation are C.reinhardii and C.eugametos . They are sexually incompatible and only distantly related to each other (Lemieux and Lemieux, 1985). It is therefore not surprising that rather obvious differences exist between them. Indeed some will be mentioned in this review of the most recent work on C.eugametos . Nonetheless, these two species have maintained or co-evolved similar strategies for sexual reproduction. Thus while research on Chlamydomonas falls into two camps, the two complement each other, for any differences tend to be quantitative rather than qualitative.

Gametogenesis and gamete flagella

C.eugametos is heterothallic and isogamous. It is cultivated as two separate mating types referred to as plus (mt^+) and minus (mt^-) usually in Petri dishes on an agar medium (Mesland, 1976). When 2-, 3-week old cultures are flooded with water, the cells swim free within a few hours as sexually competent cells. Between 80 and 100% of them will fuse when mixed with an excess of gametes of the opposite mating type. Gametes also form in liquid cultures, where they appear during the last stage of the expontential growth phase (Tomson et al. 1985). Gametogenesis is not necessarily correlated with the depletion of nitrate in the medium although it is invariably correlated with the cessation of growth and can probably be triggered by several forms of nutrient stress. It can be induced by transferring vegetatively dividing cells into low nutrient medium, but the percentage of fusion-competent cells is low (30%) and therefore liquid cultures of C.eugametos have so far found little practical application. During gametogenesis, the flagella differentiate into sex organelles that orchestrate cell fusion. The flagellar membrane, which plays the key role, is a specialised domain within the cells plasma membrane. It is continous with that of the cell body but is isolated from it by a barrier in the transition zone (Musgrave et al., 1986). Its morphology (Monk et al., 1983), composition (Musgrave et al., 1986) and its ability to transport components over its surface (Bloodgood., 1986) distinguish it from the cell membrane. The glycoproteins that epitomise the flagella as sex organelles, the cell recognition factors or agglutinins as

Algal Development (Molecular and Cellular Aspects)
Ed. W. Wiessner, D.G. Robinson, and R.C. Starr

they are called, are also present on the cell body surface (Pijst et al., 1983). They are not able to diffuse or be transported from one domain to the other (Musgrave et al., 1986) and since the agglutinins on the cell body are covered by a cell wall, they cannot partake in the initial gamete-gamete recognition. Thus only those on the naked flagellar suface are involved. Flagellar agglutination not only holds potential partners together but induces all subsequent steps leading to fusion. For example, it induces the formation of a plasma papilla, the structure by which the gamete will eventually fuse with its partner. However, the papillae must first be accurately aligned, for they only just protrude through the cell wall of the flagellar ridge. Consequently, alignment is more critical than in C.reinhardii , where the length of the mt^+ (1-2u) allows more flexibility. Again flagellar agglutination solves this problem, for the mechanics of agglutinaton automatically bring the papillae into close juxtaposition, as will be explained in a later section.

Agglutinins and monoclonal antibodies

The key to a molecular understanding of sexual agglutination is the identification of the agglutinins. Because of their visible biological activity, their presence can be readily assayed and thus the first component in the chain that activates all subsequent responses is the one most readily to hand. Both the mt^+ and the mt^- agglutinins (A^+ and A^-) of C.eugametos have now been identified (Musgrave et al., 1981; Klis et al., 1985), partially characterised and visualised. They are extrinsic membrane glycoproteins rich in hydroxyproline and serine. They are heavily glycosilated, with the sugars (Ara, Rha, Xyl, Man, Gal, Glc; mt^+-GlcNac) accounting for approximately half of the $1.2 - 1.3 \times 10^6$ molecular mass. Under the electron microscope, the A^+ resembles the A^+ and A^- of C.reinhardii (Goodenough et al., 1985) in that it is a long (290 nm), fairly rigid, rod-like molecule with a knob at one end and a hook at the other. In contrast, the A^- (320 nm) appears to be built up of five rigid sections coupled together via flexible joints. The molecules can bend back on themselves and form loops. They lack an obvious polarity but by measuring the rigid sections, one sees that they are of different lengths and a polarity is then evident.

Since the agglutinins contain about 10% hydroxyproline, we think that much of the amino acid backbone of the glycoprotein assumes a poly-proline II helix with 0.31 nm between each two peptide bonds. If the linear form of the molecules represents the length of the amino acid backbone, then the 290 nm of the A^+ can accomodate 935 amino acids. However the total protein mass is about 6×10^5 (half the molecular mass), which corresponds to about 5000 amino acids. Therefore the molecule may be composed of several peptide chains running

paralles to each other, or a single chain may bend back on itself.
An important tool for studying agglutinins are specific monoclonal antibodies (Mabs). We have several that block the acitivity of isolated A^- and that of intact mt^- gametes. Using the immunofluorescence test they can be shown to bind to gamete flagella but not to vegetative cell flagella that are non-agglutinable. In blots of gels of mt^- flagellar components, they only recognize the A^- but surprisingly, in blots of mt^+ flagellar components, they recognize at least one high molecular weight glycoprotein. This could be the A^+, but its acitivity is not affected by the antibodies. We have made use of the Mabs to visualize A^- during sexual agglutination, but because they are not mt^--specific one must somehow be able to recognize the mt^- cells. This is possible because we have selected mt^- strains that incorporate the sugars 6-O-methyl-mannose and 3-O-methyl-glucose into their surface glycoproteins. Being unusual sugars, they are very antigenic and therefore by using other Mabs one can specifically mark the mt^- cells.

Natural control of agglutinability

Once gametes have differentiated, and agglutinins have been incorporated into the flagellar membranes, one might expect them to be sexually competent, but this need not be the case. For example, the agglutinability of C.eugametos gametes is strongly influenced by light and the presence of a compatible partner. Light has two separate effects. Firstly, the light/dark regime entrains a diurnal rhythm in agglutinability in all cells. They are the most agglutinable at the end of the dark period and the beginning of the light period. Thereafter activity declines, sometimes even disappearing towards the end of the light period. In continuous light, gametes can become non-agglutiable and remain so. In continuous darkness, however, the diurnal rhythm is maintained, which indicates the presence of an endogenous rhythm . The natural fluctuations in mt^- agglutinability are correlated with the binding of anti-A^- that recognizes the mt^- agglutination site. This implies that the number of agglutinins on the flagellar membrane rhythmically fluctuates. This is confirmed by SDS-gel electroforesis of flagellar extracts, for the intendity of the A^- band fluctuates accordingly.
Apart from any negative effects that light can have on agglutinability, in some mt^+ and mt^- strains it is an absolute requirement for activity. If such strains are kept in the dark they remain non-agglutinable, but after just 15 min. in the light, maximum agglutinability is achieved. On returning to the dark, they become non-agglutinable after 30 min. This activity can be turned on and off for at least four cycles in the absence of protein synthesis, which indicates that the mechanism involves the modification/demodification of the same agglutinins rather than their turnover in the flagellar membrane (Kooijman

et al., 1986). Using a light sensitive mt^- strain, we have shown that just as much inactive mt^- agglutinin is present on flagella of dark-grown gametes as there is active agglutinin on flagella of light-grown gametes. The active form of A^- binds anti-A^- but the native inactive form does not. However, when the latter is denatured during SDS-gelelectrophoresis and Western blotting, then it binds as effectively as the active agglutinin. Thus during desactivation in the dark, the antgenic site becomes cryptic, but it can be artificially exposed by denaturing the agglutinin.

Independent of all forms of light control, the agglutinability of gametes is increased by contact with compatible sexual partners, or their isolated flagella, or their agglutinins. For example, if gametes are fixed with glutaraldehyde 10 min. after the start of sexual agglutination, and the mt^- cells used in a dilution series to agglutinate free swimming mt^+ gametes, then the titer of the fixed cells is usually 8 x higher than that of fixed, non-agglutinating cells (Tomson et al., 1986). Again the changes in mt^- flagellar agglutinability can be monitored by the binding of anti-A^-. We conclude therefore that activation involves incorporating new sites into flagellar membrane, or activating inactive sites already present. However, we have so far failed to detect more agglutination activity in extracts of activated flagella compared with non-activated flagella, therefore the explanation of this phenomenon remains open. The result of activation is that contacts between potential partners become more stable, and that the induction of subsequent events becomes stronger.

Intracellular signals and a change in motibility

Compatible gametes collide by chance (Tomson et al., 1986) and agglutinate via their flagellar surfaces. The agglutinins are extrinsic membrane proteins, therefore the fact of their mutual adhesion must be transmitted across their respective membranes, via other components, into the cells. This creates a signal that stops normal flagellar beating and triggers events that lead to fusion. Potential signals have been identified. The concentrations of cyclic AMP in *C.eugametos* (Pijst et al., 1984) and free Ca^{++} in *C.reinhardii* (Bloodgood and Levin, 1983; Kaska et al., 1985) have been shown to increase within minutes but, while both may well affect flagellar motility, no causal relationship has been shown. The rapid change in swimming is best seen when mt^+ gametes are treated with A^-, for within seconds, the gametes close to the air/water interface become immobilised at the interface although they twitch or vibrate violently. During the initial stages of agglutination, flagellar movement is too rapid to be observed but later, when partner flagella adhere over their whole length, wave movements, typical of backward swimming can be observed. Another rapid effect that is associated with a change in motility, is flagellar tip activation. It was originally described for C.reinhardii (Mesland et al,. 1980) but is just as relevant for C.eugametos (Elzenga et al., 1982). It involves the lengthening of the outer axonemal microtubules, such that the tip becomes stumpy rather than pointed. Because the ends of the microtubules are now fixed in relation to each other, this could restrict the normal sliding between tubules. In support, we have shown that the mt^- flagella of fused pairs do not beat and remain activated, while the mt^+ flagella deactivate and resume beating (Crabbendam et al., 1984).

Tipping

The initial adhesion between partner flagella can be found anywhere on the flagellar surface, yet during later stages of agglutination (or isoagglutination, Wiese, 1974; Mesland and van den Ende, 1979), some gametes can be seen adhered only by their flagellar tips. Thus it has long been possible to argue for a redistribution of agglutination sites to the flagellar tips. The work of Bloodgood and his colleagues (1986) on the transport of particles over the flagellar surface established that a mechanism for redistribution is present, but only via the use of anti-A^- Mabs has the tipping of agglutination sites been convincingly demonstrated. Before agglutination starts, A^- is evenly distributed over the whole flagellum, but during sexual agglutination, the A^- becomes concentrated at many of the flagellar tips. The redistribution takes place as a direct result of adhesion, for when cell aggregates are fixed intact, the A^- is only concentrated at the tips of those flagella seen to be lined up with their partners. What is more, when cells de-adhere after fusion, tipping also disappears. Lastly, when cells are treated with 1 mM cysteine to prevent cell fusion, gametes agglutinate for many hours, and then practically all flagella become tipped. We assume that sexually bound A^- is transported to the tips but this has not been proven, but when living mt^- gametes are treated with anti-A^- and washed, then transport of the bound Mabs becomes obvious. After 30 sec. the evenly distributed A^- is seen condensed into patches, while after 5 min. it is concentrated at the flagellar tips. Although the agglutinins are transported to the tips, the majority of surface glycoproteins are not, thus there is no mass transport. While tipping is developing, the anti-A^- antigenicity of some flagella increases in accord with the known increase in flagellar agglutinability of the whole surface which is essential for bringing the papillae into mutual contact (see next section). However, we do not understand how some but not all A^- becomes tipped.

Papillar contact

Agglutinating gametes are in continous motion, therefore partners that first adhere via two flagella have a good chance that the other two will eventually make contact. Even so, if the flagella are aligned tip-to-base (Fig. 1c), they may loose contact, for due to tipping they become attached only via their tips and that could prove too labile for the vigorous motion of the flagella. In contrast, tip-to-tip and base-to-base associations between partner flagella are stable, for tipping brings the tips into juxtaposition and maintains them there (Fig 1c), allowing further adhesions over the rest of the flagellar surface. Significantly, this aligns the papillae which just penetrate the walls of the respective flagellar ridges. Nonetheless, the papillae will not make contact while both gametes hold their flagella forward (Fig. 1a). As if to solve this problem, the mt^- flagella assume an unusual orientation, they reflex around their own cell body (Musgrave et al., 1985), and since this is complemented by the mt^+ flagella, papillar contact is made possible (Fig.1b).

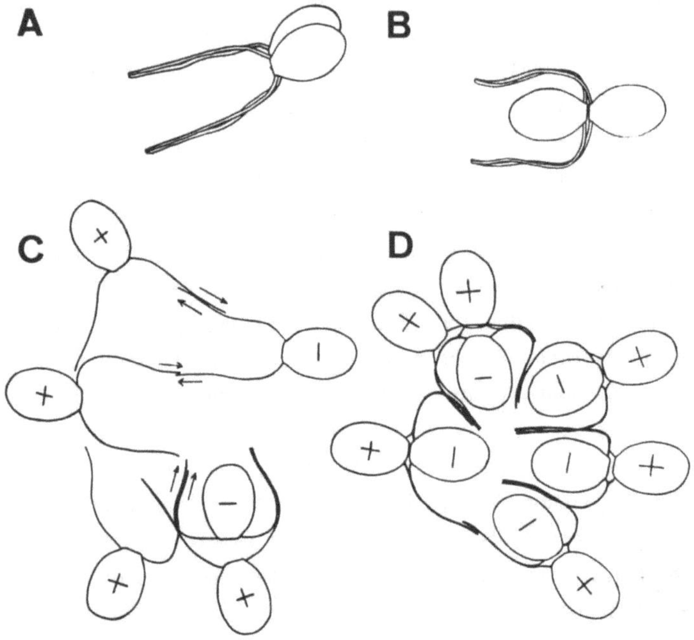

Fig.1. Diagram illustrating how tipping and flagellar orientation result in the alignment of papillae and the asymetric orientation of the mating types in clumps.

With the flagella now adhered over their whole length, motility is severely limited, even so, the partner cell bodies are continuously buffeted together such that one can imagine that the papillae are practically forced into each other. The characteristic orientation of flagella during the later stages of agglutination has a remarkable consequence for the organisation of agglutinating clumps; the mt^- gametes become confined to the interior with the mt^+ on the outside (Fig.1d). Since all the flagella of the pairing gametes are now orientated around the mt^- cell bodies, any secondary contacts made with other gametes automatically fixes the mt^- gametes within the centre of the clump. Agglutination eventually leads to cell fusion.

Acknowledgment . I thank my colleagues for their support and for permission to report their unpublished results.

References

Bloodgood, R.A. (1986) In: Advances in Cell Biology vol.1 in press.

Bloodgood, R.A., E.N. Levin (1983) J. Cell Biol. 97, 397-404.

Crabbendam, K.J., N. Nanninga, A. Musgrave, H. van den Ende (1984) Arch. Microbiol. 138, 220-223.

Elzenga, J.T.M., N. Nanninga, P.F. Lens, H. van den Ende (1982) FEMS Microbiol. L. 15, 279-283.

Goodenough, U.W., W.S. Adair, P. Collin-Osdoby, J. Heuser (1985) J. Cell Biol. 101, 824-941.

Kaska, D.D., I.C. Piscopo, A. Gibor (1985) Exp. Cell Res. 160, 371-379.

Kooijman, R., T. Elzenga, A. Musgrave, F. Schuring, H. van den Ende (1986) Planta, in press.

Lemieux, B., C. Lemieux (1985) Curr. Genet. 10, 213-219.

Mesland, D.A.M. (1976) Arch. Microbiol. 109, 31-35.

Mesland, D.A.M., H. van den Ende (1979) Protoplasma 98, 115-129.

Mesland, D.A.M., J.L. Hoffman, E. Caligor, U.W. Goodenough (1980) J. Cell Biol. 84, 599-617.

Monk, B.C., W.S. Adair, R.A. Cohen, U.W. Goodenough (1983) Planta 158, 517-533

Musgrave, A., E. van Eijk, R. te Welscher, R. Broekman, P. Lens, W. Homan, H. van den Ende (1981) Planta 153, 362-267.

Musgrave, A., P. de Wildt, K. Crabbendam, H. van den Ende (1985) Planta 166, 234-243.

Musgrave, A.,P. de Wildt, I. van Etten, H. Pijst, C. Scholma, R. Kooijman, W. Homan, H. van den Ende (1986) Planta 167, 544-553.

Pijst, H.L.A., R.J. Zilver, A. Musgrave, H. van den Ende (1983) Planta 158, 403-409.

Pijst, H.L.A., R. van Driel, P.M.W. Janssen, A. Musgrave, H. van den Ende (1984) FEBS Letters 174, 132-136.

Tomson, A.M., R. Demets, N.P. Bakker, D. Stegwee, H. van den Ende (1985) J. Gen. Microbiol. 131, 1553-1560.

Tomson, A.M., R. Demets, C.A.M. Sigon, D. Stegwee, H. van den Ende (1986) Plant Physiol. 81, 522-526.

MOLECULAR RECOGNITION PROPERTIES OF CHLAMYDOMONAS HRGP's

W. Steven Adair, Biology Department, Washington University
St. Louis, MO. 63130

Introduction

Animals and plants both possess matrix macromolecules charactererized by a high content of proline residues postranslationally modified to 4-transhydroxyproline. In animal cells, a large family of well characterized hydroxyproline-rich glycoproteins (HRGP's), the collagens, perform a variety of important structural, developmental, and homeostatic functions. That plants also contain HRGP's was first demonstrated by D.T.A. Lamport (1), who named the HYP-containing cell wall glycoprotein of dicotyledenous plants "extensin", based on a presumed role in extensibility. The insolubility of extensins, however, hampered the study of these macromolecules and their assembly properties until the discovery of soluble extensins (2-4). Like the collagens, plant extensins appear to be members of a large family of related proteins (see reference 5 for a recent review).

Much of what is known about the recognition/assembly properties of plant HRGP's comes from studies of unicellular algae like *Chlamydomonas reinhardtii*. Like other members of the Volvocales, *Chlamydomonas* has a cell wall lacking the abundant polysaccharide polymers characteristic of higher plant walls, containing instead HRGP's as major structural elements (6-8). Moreover, one layer of the *Chlamydomonas* cell wall (W6) is made up of HRGP's that are selectively extracted with a number of chaotropes and self-assemble *in vitro* (8-13). These features, plus the fact that *Chlamydomonas* is an excellent genetic system from which several cell wall mutants have been isolated (14-16), recommend this organism for the study of cell-wall HRGP assembly.

Over the past few years it has become apparent that *Chlamydomonas* has an additional class of HRGP's with well-defined molecular recognition properties. Molecules of this class, sexual agglutinins, are responsible for the mating type-specific recognition and adhesion of *Chlamydomonas* gametes, and appear to be related to cell wall HRGP's. To date, agglutinins with similar properties (functional, structural, chemical) have been described for two *Chlamydomonas*

Algal Development (Molecular and Cellular Aspects)
Ed. W. Wiessner, D.G. Robinson, and R.C. Starr

species, C. reinhardtii (reviewed in references 17, 18) and C. eugametos (see the article by Musgrave in this series). These findings raise the interesting possibility that molecules serving matrix and cell-cell recognition functions share a common evolutionary heritage in Chlamydomonas. If cell-wall and agglutinin HRGP's are related, one can predict that their recognition properties reside in homologous domains, and begin to consider the cell-cell recognition activities of agglutinins in the context of self-assembly. In the following pages I will present an overview of two groups of Chlamydomonas HRGP's, crystalline-layer (W6) cell wall glycoproteins and sexual agglutinins, currently under investigation in our laboratory. The emphasis will be on common features of these molecules that suggest that they are members of a large family of related proteins and experimental approaches currently being employed for the study of their molecular recognition properties.

Crystalline Layer HRGP's

Background

The intact cell wall of Chlamydomonas reinhardtii was originally described as having seven distinct layers, numbered W1-W7 (6). Recent images obtained using the quick-freeze deep-etch technique (19), however, indicate that W3 and W5 are actually spaces, traversed by components emanating from surrounding layers. Each of the "true" layers contains a distinct set of molecules. The innermost (W1) and outermost (W7) layers are composed of anastomosing fibers, while the central domain contains an outer crystalline layer (W6), a medial granular layer (W4), and an inner dense layer (W2). The crystalline layer is further differentiated into inner (W6A) and outer (W6B) sublayers, each containing unique molecular species.

In a series of classic papers (reviewed in reference 8) Keith Roberts and coworkers demonstrated that a subpopulation of cell wall components, extracted by chaotropic agents (e.g. sodium perchlorate), reassemble in vitro when the chaotrope is removed by dialysis. They further showed that the reassembled material had a lattice structure characteristic of the crystalline layer (W6) of the intact cell wall (6-8,11). Recently, Goodenough et al. (20) reinvestigated this phenomenon and showed that : (1) crystal formation is highly concentration-dependent; and (2) individual crystals are sandwiches of

W6-W4-W6 layers. Moreover, by taking advantage of the concentration dependence of crystal assembly, they were able to purify and characterize each of the molecular species, and show that W6 crystals can assemble _in vitro_ in the absence of W4. While GP-1.5, the globular molecule that makes up W4, was found to have little hydroxyproline (but significant amounts of glycine), each of the W6 components (GP-1, GP-2, GP-3) was shown to be rich in this amino acid. Thus, the crystalline layer is made up entirely of HRGP's that contain all the information required for assembly of the W6 matrix. Two of these (GP-2 and GP-3) have been assigned to the inner (W6A) sublayer, while a single glycoprotein (GP-2) self-assembles to form the outer (W6B) sublayer (20).

GP-1 corresponds to the fibrous 100nm species described by Roberts (21) and Goodenough and Heuser (19). It has a prominent globular head at one end and has bends at two characteristic positions (20). Of the three W6 molecules, GP-1 is by far the richest in hydroxyproline (>30%). As detailed in a later section, the globular head of GP-1, which may participate in self-assembly into the W6B weave, bears a carbohydrate epitope shared by the globular head and hook domains of the _Chlamydomonas_ _plus_ agglutinin (17). These agglutinin domains may also be endowed with molecular recognition properties (22,23).

The GP-2 and GP-3 molecules are much less fibrous than GP-1 and have similar but distinctive morphologies (20). Each has a small terminal head, followed by a fibrous domain and two (GP-2) or three (GP-3) globular domains. GP-2 is moderately rich in hydroxyproline (approx. 15%), while GP-3 is much less so (approx. 6%). In addition, GP-2 is a monomer, while GP-3 dissociates into two subunits (GP-3A and GP-3B) in the presence of reducing agents. Thus, the two components that coassemble to form the W6A sublayer are distinct, but possibly related, molecular species.

Nucleated Assembly of W6 Components

In their original studies of the _Chlamydomonas reinhardtii_ cell wall, Roberts and coworkers extracted crystalline layer (W6) components from walls and wall fragments shed into the growth medium, using 8M LiCl or 1M sodium perchlorate. They observed that, while the soluble components alone were capable of assembling into crystals, the insoluble residue had nucleating activity (24). Thus, component(s) of the inner wall contain specific binding sites for crystalline layer glycoprotein(s). This conclusion is consistent with the observation that the cell wall mutant cw-2 synthesizes apparently normal W6 components, but releases them into the medium (14), implying a defect in the inner layer. Based on this information and the recent demonstration that whole cells, extracted with perchlorate, remain intact (19), we have developed the following experimental system to study the individual steps of W6 assembly (Adair, Steinmetz, Mattson, Goodenough, and Heuser, submitted for publication).

For these studies, perchlorate-extracted cells are employed as substrates for the nucleated assembly of crystalline layer glycoproteins. Following extraction, cells are fixed with paraformaldehyde, followed by incubation with buffered glycine to block remaining free aldehydes. When prepared in this fashion, cells retain their nucleating ability for weeks at 4°C. Crystalline components, isolated from bald-2 cells as described previously (19,20), are biotinylated using a N-hydroxysuccinimide derivative of biotin (NHS-biotin) to facilitate monitoring of reassembly by FITC-streptavidin fluorescence microscopy. Reassembly is initiated by mixing extracted cells and biotinylated crystals in the presence of 1M perchlorate and dialyzing the mixture against distilled water. To facilitate the analysis of several samples at once, a multiple dialysis manifold (BRL) is employed. Alternatively, biotinylated W6 components are presented to cells directly, at a concentration below that required for crystal self-assembly (20). Identical results are obtained using both procedures.

To define optimal conditions for nucleated assembly, cells were incubated with increasing concentrations of W6 monomers and examined by fluorescence microscopy after reaction with FITC-streptavidin. Under the conditions employed, optimal wall occurred at a total protein concentration of 0.3-0.5mg/ml, well below that required for initiation of crystal self-assembly (20).

To confirm that an organized matrix was assembled, duplicate samples were examined by electron microscopy using the quick-freeze deep-etch technique (25). Images obtained clearly demonstrated the presence of W6 components, organized into a typical crystalline structure, and indicated that reassembly occurs from multiple nucleation sites: reassembled W6 components appeared in patches, each displaying an apparently random orientation with respect to its neighbors. During *in vivo* assembly, therefore, the cell must provide a polarizing influence (e.g. secretion) that establishes the orientation of W6.

The specificity of nucleated assembly was confirmed by two lines of evidence. First, no assembly of crystal components was obtained when the cell wall mutants cw-2 and cw-15 were employed as substrates. W6 components released into the medium by cw-2 were, however, able to assemble onto wild-type cells, supporting the notion (15) that cw-2 has an inner wall defect. The second line of evidence comes from interspecific reconstitution experiments. Roberts and coworkers (reviewed in reference 8) have examined a large number of species within the Volvocales, and have found that their crystal structures fall into four general classes. C. eugametos, a distantly related species of C. reinhardtii, possesses the simplest crystal structure, typical of most of these organisms. Roberts (8) has speculated that this structure may be the closest present-day descendant of a hypothetical ancestral wall. C. reinhardtii has a cell wall crystal structure more typical of the multicellular, colonial members of the Volvocales, including Volvox (26). When crystal extracts were prepared from C. eugametos and Volvox carterii, each preparation was capable of assembling onto the substrate of origin. No reconstitution was observed, however, when C. eugametos and C. reinhardtii cells were incubated with crystal components of the other species. Thus, nucleated assembly requires specific association with the underlying matrix. That this association is not rigorously species-specific, however, was indicated by the finding that successful interspecific reconstitutions were obtained between C. reinhardtii and Volvox, in both combinations. These findings support the notion that C. reinhardtii is on the evolutionary line leading to the multicellular Volvocales, and indicate a degree of molecular conservation within this line.

As detailed above, the W6 layer is made up of inner (W6A) and outer (W6B) sublayers containing unique molecular species. W6A is formed by the coassembly of two glycoproteins, GP-2 and GP-3, while a single glycoprotein, GP-1, self-assembles into the open weave of W6B. As a first step in the analysis of W6 assembly reactions, we have determined that purified GP-2 and GP-3 components will coassemble onto extracted cells in vitro to form W6A sublayer. Significantly, GP-1 will not assemble onto such a substrate. GP-1 will, however, assemble a W6B sublayer onto cells containing a reconstituted W6A. Taken together, these results indicate that the inner cell wall of C.

reinhardtii has specific binding sites for GP-2 and/or GP-3 and that the coassembly of these macromolecules into W6A is required for the self-assembly of GP-1.

Chlamydomonas Sexual Agglutinins

The Mating Reaction

Chlamydomonas cells differentiate into isogamous plus and minus gametes in response to nitrogen deprivation (27), a process requiring approximately eight hours and accompanied by a round of mitosis. During gametic differentiation, cells acquire on their flagellar surfaces molecules (agglutinins) that are responsible for mating type-specific agglutination. When plus and minus gametes are mixed, agglutination is instantaneous, resulting in large clumps of cells tethered by their flagella. As a consequence of flagellar agglutination signals are generated that prepare gametes for zygotic fusion. One signal elicits the release of a wall lytic factor (autolysin), while a second signal leads to activation of mating structures, specialized regions of the plasmalemma that engage in mating type-specific fusion. Fusions of mating pairs results in the rapid loss of flagellar agglutinability and release of the quadriflagellated zygote from mating clusters. This has the important consequence that mating efficiency can approach one hundred percent.

Isolation of Agglutinins

Early studies by Lutz Wiese and coworkers established that flagellar adhesion is a highly specific process, mediated by flagellar glycoproteins (28,29). Studies using fixed cells (30), isolated flagella (20), membrane-derived vesicles (31,32), detergent extracts of isolated flagella (33) and EDTA extracts of living gametes (33) showed that both protein and carbohydrate components are important for agglutinin function. These results have been confirmed by studies with purified agglutinin preparations, immobilized on a solid support (34), showing that agglutinin activity is destroyed by a variety of proteases, periodate oxidation, and β-elimination.

Identification of the plus (35) and minus (23) agglutinins was made possible by the development of a sensitive, quantitative in vitro bioassay (33). For this procedure, agglutinin-containing extracts were prepared from isolated flagella, using the dialyzable detergent octylglucoside, or from whole cells, using EDTA. When a small drop (1 µl) of this material was applied to a microscope slide, dried and presented to gametes of the opposite mating type, a strong agglutination response was observed. The specificity of the response (gametes of the same mating type showed no reaction) confirmed that agglutinin activity was being assayed. This conclusion was further supported by the demonstration that responding cells activate their standard fertilization program. Moreover, extracts prepared from non-agglutinating cells (vegetative cells and agglutination-deficient mutants) were inactive.

The plus agglutinin was initially isolated from ^{125}I surface-labeled gametes by gel filtration on Sepharose 6B and Fractogel F-75 columns (35). Fractions were assayed for agglutinin activity by quantitative bioassay and their constituent polypeptides identified by SDS-PAGE/autoradiography. When examined in this fashion, agglutinin activity consistently co-fractionated with a high molecular weight glycoprotein, termed "polypeptide A". The identity of "polypeptide A" as the plus agglutinin was confirmed by examination of ^{125}I extracts prepared from sag-1 and sag-2 mutants and polyethylene glycol (PEG)-induced diploids. sag-1 mutants (imp-2:imp-5), and the sag-2 mutant (imp-8) all were found to lack "polypeptide A", as did non-complementing diploids (e.g. imp-2, imp-5). In contrast, complementing diploids, created by fusion of imp-5 (sag-1) and imp-8 (sag-2), had restored agglutination function and displayed "polypeptide A". Based on these biochemical and genetic data, it was concluded that "polypeptide A" is the Chlamydomonas plus agglutinin. Similar procedures were subsequently employed to identify and purify the minus agglutinin (23). As detailed below, the two molecules are homologous by several criteria.

While lacking a molecule corresponding to "polypeptide A", several non-agglutinating mutants were found to possess a faster migrating component, termed "band B". Subsequent morphological analysis (22) revealed that "band B" contains two major species (short canes and loops). Remarkably, these two molecules bear a strong resemblance (structural and immunological) to the two halves of the agglutinin molecule. This observation, together with the finding that extracts from wild-type gametes contain a subpopulation of imp-like molecules, raises the possibility that the active agglutinin is a conjugate of two separate gene products. To date, however, a precursor-product relationship between short canes and loops has not been demonstrated.

Characterization of Agglutinins

Plus and *minus* agglutinins fractionate on gel permeation columns in a manner characteristic of highly asymmetric proteins. When visualized using the quick-freeze deep-etch technique, both molecules are found to be extended, rod-like proteins (225nm x 4nm) possessing several prominent structural landmarks (22,36). Each has a globular head at one end, a hook region at the opposite end, and a rod-like shaft. While clearly homologous, *plus* and *minus* agglutinins are easily distinguished by the conformation of the domain immediately adjacent to their globular heads. In the case of the *plus* agglutinin, this region invariably appears straight, while the corresponding *minus* domain appears curved.

Based on the solubility and extraction properties of agglutinins, it was concluded that these molecules associate with the flagellar surface in an extrinsic fashion (33). *In situ* visualization of agglutinins demonstrated that agglutinins attach to the flagellar membrane via their hook end, projecting their globular domain distally (22). Thus, the hook end is identified as a binding domain. That the head is important for agglutinin function is indicated by two lines of evidence. First, thermolysin digestion, which inactivates both agglutinins (34), has its primary effect on the head (22). Secondly, reduction and alkylation, which inactivates the *plus* agglutinin (34), converts its head into an open structure (22), indicating that the globular morphology of this domain involves intermolecular disulfide bond(s). A similar analysis has not been completed with the *minus* agglutinin; however, it should be noted that its activity is much more resistant to reducing agents (32).

The first indication that *Chlamydomonas* agglutinins are HRGP's came from the amino acid analysis of purified *plus* agglutinin (36), which was found to contain approximately 12% hydroxyproline and 10% serine. Similar values were subsequently obtained for the *minus* agglutinin (23). As noted elsewhere (36), agglutinin compositions resemble those reported by Roberts and collaborators for the major cell wall glycoproteins of *Chlamydomonas* (10).

That proline hydroxylation is required for agglutination is demonstrated using the _plus_ mutant _imp-1_ which agglutinates normally but is unable to fuse. In the absence of zygotic fusion, _imp-1_ gametes continue to agglutinate with _minus_ gametes for hours. In the presence of cycloheximide, however, the cells come apart within 20 minutes (38), implying that continued agglutination requires a continual replacement of agglutinin molecules via _de novo_ synthesis. When similar experiments are carried out using inhibitors of proline hydroxylation (α,α'-dipyridyl and 3,4 dehydroproline), cells are found to disadhere with approximately the same kinetics obtained with cycloheximide (36). From the recovery kinetics of α,α'-dipyridyl inhibition in the presence of cycloheximide, it appears that a nonhydroxylated precursor is synthesized in the presence of the drug, which can serve as a substrate for hydroxylase once inhibition is relieved. Thus, proline hydroxylation is required for expression of agglutination activity. Whether the requirement involves agglutinin processing and/or secretion has not been determined. From analogy with cell wall HRGP's, however, one might predict that proline hydroxylation and associated arabinosylation are required for the structural integrity of agglutinins.

Important information on the organization of agglutinin molecules is obtained by high resolution epitope-mapping, using a library of monoclonal antibodies generated against the _plus_ agglutinin (39), and the quick-freeze deep-etch technique to visualize bound antibodies. When analyzed in this fashion, the immunodominant agglutinin epitope (Class I) is localized to the two ends (head and hook) of the _plus_ agglutinin; no binding to the rod is observed. As noted above, both domains are believed to have recognition function.

The Class I epitope is sensitive to mild periodate oxidation and recognizes many polypeptides, including cell wall glycoproteins, when examined by gel overlay analysis (39). These properties suggest that the epitope is a carbohydrate; however its chemical nature has not been further defined. It is perhaps significant that Class I antibodies bind to a corresponding site on the cell wall glycoprotein GP-1 (unpublished data from this laboratory). Like the agglutinins, GP-1 is an extended HRGP, with a globular head at one end. GP-1, however, is shorter (100nm) than the agglutinins, and lacks a terminal hook. Class I antibodies bind only to the head of GP-1, a site that is apparently endowed with recognition properties.

Two additional periodate-sensitive epitopes have been mapped on the plus agglutinin. One of these (Class 3), is also localized to the ends of agglutinin, but displays a unique crossreactivity with the major flagellar membrane glycoprotein (350kD) of Chlamydomonas. The second (Class 4), while not completely characterized, is present in a repeating fashion along the shaft of the plus agglutinin.

To date, a single periodate-insensitive epitope has been mapped on the plus agglutinin. This epitope (Class 2) is agglutinin (but not mating type)-specific. Like the Class 4 epitope, it is displayed in a repeating fashion along the shaft of the plus agglutinin. In summary, epitope-mapping studies indicate that the plus agglutinin contains: (1) at least two distinct carbohydrate determinants common to the two ends (head and hook) of the molecule, that are absent from the shaft; (2) a shaft possessing two repeating epitopes, one periodate-sensitive and one periodate-insensitive.

Perspectives

Chlamydomonas reinhardtii has two classes of HRGP's, W6 layer cell wall glycoproteins and sexual agglutinins, whose molecular recognition properties are well documented. It is our working hypothesis that these molecules are members of a large family of HRGP's that share a common evolutionary heritage, both within the species and throughout the Volvocales. This hypothesis predicts the existence of homologous HRGP's within the Volvocales and a conservation of domains endowed with molecular recognition properties. Moreover, the inclusion of sexual agglutinins in this family suggests that, in Chlamydomonas, cell-cell recognition and matrix assembly events are mediated by similar molecular domains.

That homologous HRGP's are present within the Volvocales is implied from studies by Roberts and coworkers demonstrating a high degree of conservation in cell wall organization within the group (8). Direct evidence for homologous glycoproteins comes from recent comparative studies of C. reinhardtii, C. eugametos, Gonium pectorale, and Volvox carterii. Each species has a crystalline layer made up of a small number of high molecular weight glycoproteins. Off-diagonal peptide mapping of each preparation shows that each species has a major component corresponding to C. reinhardtii GP-2 (Appel and Adair, unpublished). In the case of C. eugametos, the GP-2 equivalent is the only major component of the crystalline layer (40).

While few details are known about the recognition domains of Volvocales HRGP's, it is useful to categorize the kinds of molecular associations that must exist. As noted earlier, isolated C. reinhardtii W6 glycoproteins self-assemble in vitro into a bilaminar structure whose lattice structure faithfully reflects the organization of the intact cell wall. All the information for the in vivo organization of W6, therefore, must reside in the molecules themselves. For assembly of the W6A layer, three distinct recognition activities can be envisaged: (1) heterotypic associations between GP-2 and GP-3; (2) homotypic associations (GP-2/GP-2, GP-3/GP-3); and (3) interactions with the underlying matrix. Assembly of the outer (W6B) layer clearly requires two additional recognition events: (1) homotypic assembly of GP-1 monomers; and (2) specific association with the W6A layer. The assembly of the C. eugametos crystalline layer, by contrast, can be accomplished by homotypic recognition (self-assembly) of what may be a GP-2 equivalent, and binding to the inner wall layer. If one accepts the idea (8) that the C. eugametos wall structure is most like the ancestral Volvocales wall, one can ask whether the other crystalline layer HRPG's (and sexual agglutinins) owe their origin to an ancestral GP-2 gene. As a first step toward answering this question, we have recently employed a polyclonal antibody to C. reinhardtii GP-2 to screen a λgt11 genomic expression library (41), and have isolated several positive clones (Adair and Mattson, unpublished). Preliminary results suggest that one of the inserts codes for an epitope common to many cell wall proteins.

References

1. Lamport, D.T.A., and D.H. Northcote. 1960. Nature 188:665-666.
2. Stuart, D.A., and J.E. Varner. 1980. Plant Physiol. 66:787-792.
3. van Holst, G.J. and J.E. Varner. 1984. Plant Physiol. 74:247-251.
4. Cooper, J.B., and J.E. Varner. 1983. Biochem. Biophys. Res. Comm. 112:161-167.
5. Cooper, J.B., J.A. Chen, and J.E. Varner. 1984. In: Structure, Function, and Biosynthesis of Plant Cell Walls, W.M. Dugger and S. Bartnicki-Garcia, eds. pp. 75-88. American Society of Plant Physiologists, Rockville, MD.
6. Roberts, K., M. Gurney-Smith, and G.J. Hills. 1972. J. Ultrastruc. Res. 40:599-613.
7. Hills, G.J., M. Gurney-Smith, and K. Roberts. 1973. J. Ultrastruc. Res. 43:179-192.
8. Roberts, K., C. Grief, F.J. Hills, and P.J. Shaw. 1985. J. Cell Sci. suppl. 2 105-127
9. Catt, J.W., G.J. Hills, and K. Roberts. 1976. Planta 131:165-171.
10. Catt, J.W., G.J. Hills, and K. Roberts. 1978. Planta 138:91-98.
11. Hills, G.J., J.M. Phillips, M.R. Gay, and K. Roberts. 1975. J. Mol. Biol. 96:431-441.

12. Roberts, K. 1979. Planta 146:275-279.
13. Roberts, K., M.R. Gay, and G.J. Hills. 1980. Physiologia Pl. 49:421-424.
14. Davies, D.R. and A. Plaskitt. 1971. Genet. Res. 17:33-43.
15. Davies, D.R. 1972. Mol. Gen. Genetics 115:3334-348.
16. Davies, D.R. 1973. Mol. Gen. Genetics 124:21-33.
17. Adair, W.S. 1985. J. Cell Sci. Suppl.2 233-260.
18. Goodenough, U.W., W.S. Adair, P. Collin-Osdoby, and J.E. Heuser. 1985. In: The Cell in Contact. G.M. Edelman, and J. Thiery, eds. pp. 111-135. Wiley, New York.
19. Goodenough, U.W., and J.E. Heuser. 1985. J. Cell Biol. 101: 1550-1568.
20. Goodenough, U.W., B. Gebhart, R.P. Mecham, and J.E. Heuser. 1986. J. Cell Biol. 103:405-417.
21. Roberts, K. 1981. Micron. 12:185-186.
22. Goodenough, U.W., W.S. Adair, P. Collin-Osdoby, and J.E. Heuser. 1985. J. Cell Biol. 101:924-941.
23. Collin-Osdoby, P., and W.S. Adair. 1985. J. Cell Biol. 101:1144-1152.
24. Hills, G.J. 1973. Planta 115:17-23.
25. Heuser, J.E. 1983. J. Mol. Biol. 169:155-195.
26. Kirk, D.L., R. Birchem, and N. King. 1986. J. Cell Sci. 80:207-231.
27. Sager, R., and S. Granick. 1954. J. Gen. Physiol. 37:729-742.
28. Wiese, L. 1969. In: Fertilization: Comparative Morphology, Biology and Biochemistry. C.B. Metz and A. Monroy, eds. Academic Press, new York.
29. Weise, L. and W. Wiese. 1978. Symp. Soc. Exp. Biol. 32:83-104.
30. Goodenough, U.W., and D. Jurivich. 1978. J. Cell Biol. 97:680-693.
31. Bergman, K., U.W. Goodenough, D.A. Goodenough, J. Jawitz, and H. Martin. 1975. J. Cell Biol. 68:48-69.
32. Snell, W.J. 1976. J. Cell Biol. 68:48-69.
33. Adair, W.S., B.C. Monk, R. Cohen, and U.W. Goodenough. 1982. J. Biol. Chem. 257:4593-4602.
34. Collin-Osdoby, P., W.S. Adair, and U.W. Goodenough. 1984. Expl. Cell Res. 150:282-291.
35. Adair, W.S., C.J. Hwang, and U.W. Goodenough. 1983. Cell 33:183-193.
36. Cooper, J.B., W.S. Adair, R.P. Mecham, J.E. Heuser, and U.W. Goodenough. 1983. Proc. Natl. Acad. Sci. USA 80:5898-5901.
37. Goodenough, U.W. 1986. Exp. Cell Res. 166:237-246.
38. Snell, W.J., and W.S. Moore. 1980. J. Cell Biol. 84:203-210.
39. Adair, W.S., J. Long, W.B. Mehard, J.E. Heuser, and U.W. Goodenough. 1983. J. Cell Biol.97:93a.
40. Musgrave, A., P. de Wildt, R. Bockman, and H. van den Ende. 1983. Planta 158:82-89.
41. Williams, B.D., D.R. Mitchell, and J.L. Rosenbaum. 1986. J. Cell Biol. 103:1-11.

MATURATION OF A FLAGELLUM/BASAL BODY REQUIRES MORE THAN ONE CELL CYCLE IN ALGAL FLAGELLATES: STUDIES ON *NEPHROSELMIS OLIVACEA* (PRASINOPHYCEAE)

M. Melkonian, I.B. Reize and H.R. Preisig*

Botanisches Institut, Universität Münster, Schloßgarten 3, D-4400 Münster, FRG and *Institut für Systematische Botanik der Universität Zürich, Zollikerstraße 107, CH-8008 Zürich, Switzerland

INTRODUCTION

Most algal flagellates bear two flagella per cell which differ from each other in size, structure and function (review: Moestrup 1982). In the green algae (Chlorophyta), however, most of the flagellated taxa bear structurally identical flagella (2,4 or rarely 8 per cell). But even in this group of algae recent evidence suggests that at least two flagella of a cell are functionally different, a prerequisite for essential behavioural responses like phototaxis (Kamiya et al. 1984; review: Melkonian et al. 1984).

What is the developmental basis for the simultaneous presence of two different forms of one cell organelle in a unicellular organism? It is well known that flagella (and/or basal bodies) are distributed among daughter cells in a semi-conservative fashion, i.e. daughter cells receive one parental and one newly formed flagellum (and/or basal body) during cell division. The two flagella (basal bodies) of a daughter cell are thus dissimilar in their developmental status. If the "maturation" time of a flagellum (basal body) requires more than one cell cycle and if different structural or functional properties are expressed during flagellar maturation, the cell would bear two structurally or functionally different flagella at the same time. This concept implies that a single flagellum changes during its development from one structural or functional form to the other. As an alternative

Algal Development (Molecular and Cellular Aspects)
Ed. W. Wiessner, D.G. Robinson, and R.C. Starr

hypothesis one may postulate that the two flagella of a cell are intrinsically different and that this applies also to the two newly formed daughter flagella (basal bodies) which are of either form. In a complex segregation pattern the cell would have to ensure that the correct parental flagellum (basal body) is distributed together with the correct daughter flagellum (basal body) to the new daughter cells. In this concept the two dissimilar flagella of a cell are not developmentally related and each flagellum completes its full development during one cell cycle as do other cell organelles.

To our knowledge the question of the developmental origin of dissimilar flagella in flagellates has not been addressed before although it has wide ranging implications for algal development and evolution.

In this study we have chosen the scaly green flagellate *Nephroselmis olivacea* to study the development of individual flagella in an organism that bears two flagella of significantly unequal length and of different motile behaviour. Our results demonstrate that maturation of a flagellum requires at least two (possibly three) complete cell cycles. Flagella that are newly formed at the beginning of the first cell cycle change to the other flagellar type during the following cell cycle. We propose that this scheme of flagellar (basal body) development is used by all algal flagellates to achieve flagellar heterogeneity in a single cell.

MATERIALS AND METHODS

Nephroselmis olivacea Stein (Prasinophyceae) was collected as the dominant organism from a small pond near the "Thalwiler Waldweiher", Canton of Zurich, Switzerland, on Dec. 10th 1985. Unialgal cultures were established using Pringsheim's biphasic soil-and-water medium (Pringsheim 1946). For experiments algae were grown in a modified Waris-solution as previously described (McFadden et al. 1986). For investigations of flagellar behaviour during the cell cycle aliquots from actively growing cultures were taken at 3hr-intervals. The length of flagella (both flagella of a cell were always evaluated) and the cell dimensions (width, length) were measured from 25-50 cells per sample as previously described (McFadden et al. 1986). Experimental flagellar amputation was performed as previously described (McFadden et al. 1986). Deflagellated cells were allowed to regenerate flagella in aerated, 30ml tubes at 20°C in the light or dark. During different time intervals of flagellar regeneration samples were taken and the length of flagella of 25 cells (again only pairs of flagella were evaluated) measured as previously described (McFadden et al. 1986). Cells were photographed for light microscopy using Nomarski-optics on a Zeiss IM35 inverted microscope with a 63x oil immersion objective.

RESULTS

The taxonomy and ultrastructure of *Nephroselmis olivacea* has been previously studied by Moestrup et al. (1979) and the reader is referred to this work for further information. Cells of *N. olivacea* are broadly bean-shaped and are compressed from the sides resulting in a pronounced difference in width between the broad and narrow side (Figs 1,2). When viewed from the broad side the two flagella can be seen to emerge from a shallow depression which gives the cell its bean-shaped outline (Fig. 6). The flagella are of considerably unequal length, the long flagellum usually at least 1.5 times the length of the short flagellum (18-25 µm vs. 12-16 µm; Figs 1,6) . Both flagella are bent backwards and during swimming the short flagellum is directed toward the direction of cell movement and beats in a cilia-type mode, whereas the longer flagellum trails and performs undulating waves passing from base to tip. The organism thus swims "sidewards". The eyespot is located at the anterior edge of the chloroplast under the short flagellum. The most conspicuous part of the cell in the light microscope

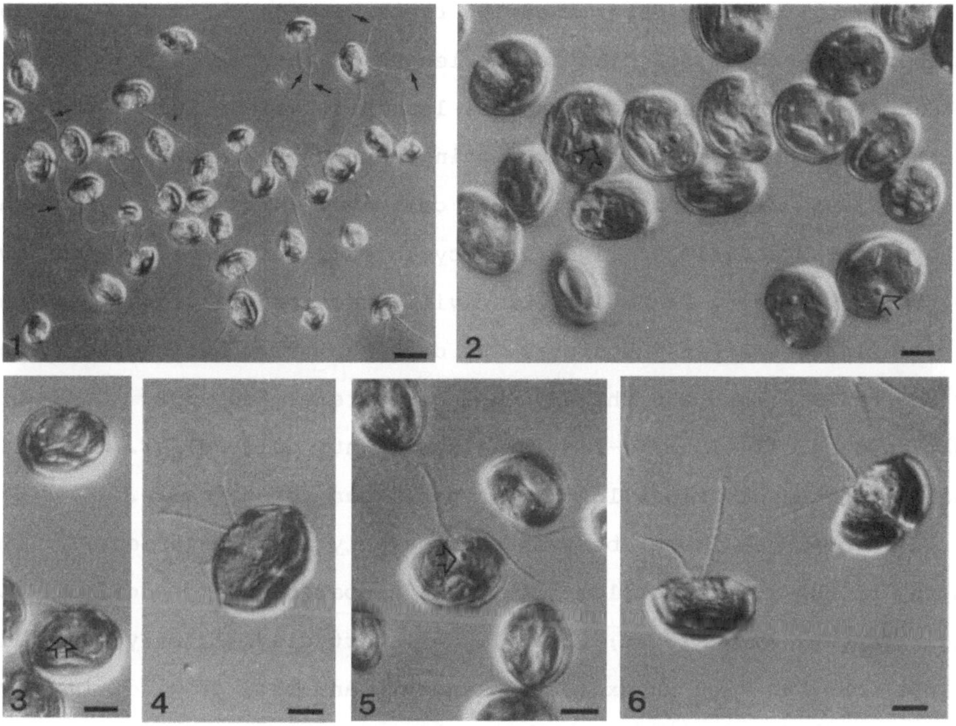

Figs 1-6 Light micrographs (Nomarski-optics) of N. olivacea. Fig.1 Overview of actively growing population. Each cell bears two flagella of considerably unequal length (arrows). Magnification bar: 10 µm. Fig. 2 Cells immediately after experimental flagellar amputation. The nucleolus (open arrows) serves as a positional marker for identification of long vs. short flagellum. Figs 2-6 Magnification bar: 5 µm. Figs 3-6 Different stages of flagellar regeneration; 1.5 hrs (Fig. 3), 4 hrs (Fig. 4), 5 hrs (Fig. 5), 8 hrs (Fig. 6) after deflagellation. Notice that the regenerated flagella are of unequal length, the longer flagellum being associated with the broad side carrying the nucleolus (Figs 3,5; open arrows).

is the starch grain-pyrenoid complex that changes size and shape during growth of a cell (Figs 3-5).

Cultures of N. olivacea cannot at present be synchronized. It is therefore necessary to identify stages in the cell cycle by morphological criteria of individual cells. Since cells primarily grow by growth in the direction of the broad cell side and not by growth in the direction of the cell axis (extending from flagellar insertion to

the pyrenoid) we have determined the ratio between the square of the length of the broad cell side to the length of the cell axis (B^2/L). A good correlation exists between this length relationship and the size and shape of the principal starch grain and thus this numerical figure seems to be a useful indicator of the cell age. In an actively growing culture using a 14/10 hr light/dark cycle each cell divides into two daughter cells every 24 hrs. Cell division occurs exclusively in the dark (during the first 6 hours of the dark period). During the light period cells grow and the B^2/L-values increase from 9-13 (0-3 hrs light; young cells) to 19-24 (12-14 hrs light; old cells). Old cells usually have longer flagella (both types) than young cells. However, in the young cells we observed considerably more variation in the length of the long flagellum (16-23 µm) compared to old cells (flagellar length: 19.5 -23 µm; Fig. 7). The short flagella of young cells do not vary to such an extent in length, and they grow during the light period from 10-11 µm in young cells to 14-15 µm in old cells.
From these initial observations it was clear that a significant proportion of flagellar growth must take place during the dark period when cells divide. To investigate the behaviour of individual flagella during cell division we measured flagellar lengths in dividing cells (quadriflagellated cells; all four flagella of a cell were evaluated) and in the youngest division products. The average B^2/L-value of quadriflagellated cells (40 cells evaluated) is 23. Dividing cells grow two additional flagella which reach a maximum length of about 9.5 µm before cytokinesis takes place. Both new flagella grow with the same velocity (ca. 50 nm/min.). During growth of the new flagella the old flagella do not change length (average 22.23 µm and 15.77 µm respectively). The youngest division products (daughter cells) appear in the cultures at the third hour of the dark period. These cells can be easily recognized by their small B^2/L-values (less than 11) and the asymmetric shape of the posterior starch grain.

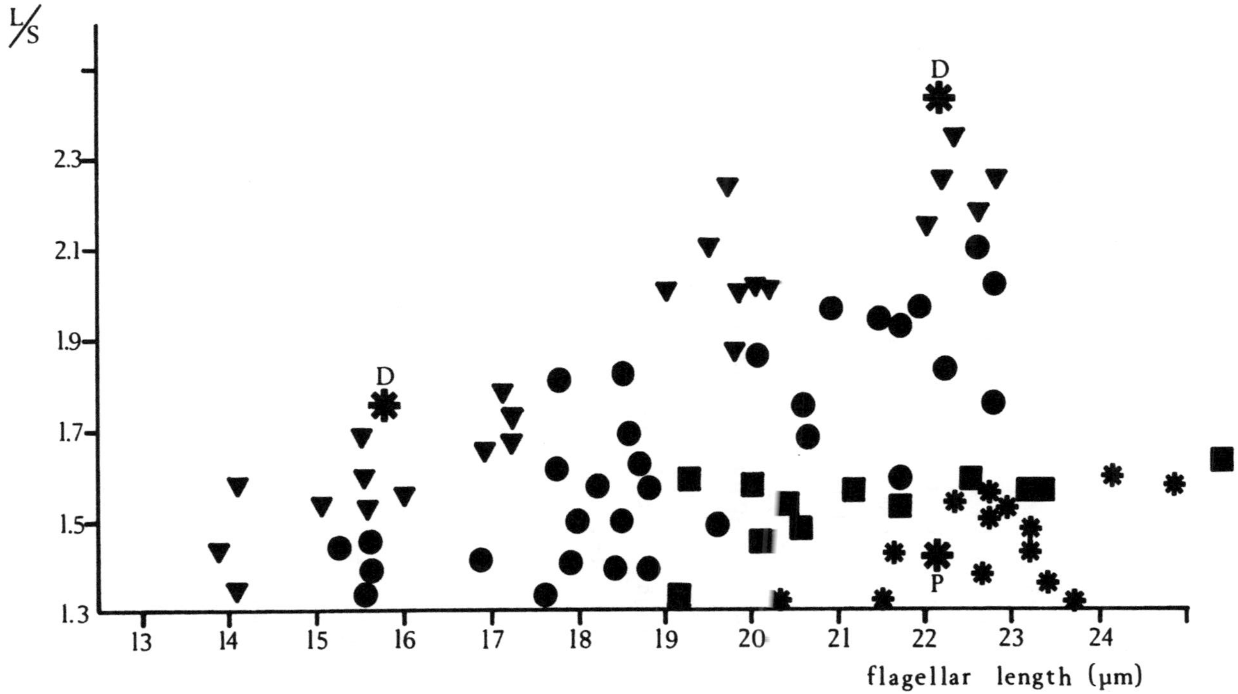

Fig. 7 Flagellar length relationships during the cell cycle of Nephroselmis olivacea. Each symbol represents the relationship between the length of the long flagellum and the ratio between the long and short flagellum (L/S) of a single cell. ✱ : quadriflagellated, dividing cells; ✱ with P : average quadriflagellated cell (n=20), ✱ with D : the same average quadriflagellated cell, but this time not the length relationship of the old flagella to each other (as with P), but between each of the old and one newly formed flagellum is plotted. ▼ : youngest daughter cells (newly formed flagella less than 10 µm); ● : young cells (BxB/L: up to 13); ■ : older cells (BxB/L: 13-20).

All these young daughter cells have short flagella of less than 11 um, the shortest being in the range of 9.5 to 10 µm. If one determines the length of the long flagellum in daughter cells with short flagella of less than 10 µm (24 cells evaluated), there exist three distinct size classes of long flagella. Size class I ranges from 14 to 17.5 µm and occurs in about 50 % of these young cells, size class II ranges from 19 to 20.5 µm and occurs in about 25 % of the cells, and size class III ranges from 22 to 23 µm and occurs in the rest of the cells. Correspondingly the length ratio between the long and short flagellum in these cells is quite variable (between 2.3:1 to 1.5:1; Fig. 7). If the length of the long flagellum is plotted against the length ratio of the long to the short flagellum (L/S-ratio), the three groups of youngest daughter cells appear as three distinct clusters (Fig. 7). If older cells are added to this graph (young cells with B^2/L-values up to 13; cells with B^2/L-values up to 20, and quadriflagellated cells), the three clusters of daughter cells merge until in dividing cells the long flagella range between 19.5 and 24 µm and the L/S-ratio is about 1.5:1 (Fig. 7). These data prove that the distribution of flagella to daughter cells follows the semi-conservative mode. 50 % of the daughter cells receive the old short flagellum which represents the new long flagellum of the young daughter cell (size class I long flagella) since the L/S-ratio at no stage of the cell cycle drops below 1.4-1.5:1 (Fig. 7). These new long flagella are still relatively short in the youngest daughter cells but they resume growth following cytokinesis (Fig. 7). The other 50 % of the daughter cells receive the old long flagellum which remains the long flagellum of a flagellar pair during subsequent growth of these daughter cells (Fig. 7). About 50 % of these old long flagella (i.e. size class II long flagella) grow for another 2-3 µm until they reach their mature length (size class III long flagella; Fig. 7). Size class III long flagella do not grow in a measurable way (Fig. 7). The occurrence of size class II long flagella in daughter cells can be explained if size class I long flagella (i.e. the short flagella of the previous cell generation) grow during interphase to size class II long flagella (19-20.5 µm) before cells begin to divide. That this may actually be the case can be deduced from the range of variation in the length of the long flagellum (19-24 µm) which persists from the middle of the light period (B^2/L: 13-20) until cell division (quadriflagellated cells; Fig. 7). This also suggests that most of the new growth of size class I long flagella takes place up to the middle of the light period. Since the L/S-ratio of cells with size class I long flagella does not change significantly during the light period (Fig. 7), the newly formed short flagella also presumably cease growth at the middle of the light period. Each flagellum of N. olivacea therefore grows during three successive cell cycles until it reaches its final length. Growth of a flagellum is discontinuous and occurs in decelerating waves: the most rapid growth takes place when a flagellum is newly formed shortly before cell division (50 nm/min) and continues after cytokinesis with slower velocity until the middle of the following light period when the flagellum reaches about 15 um length and represents the short flagellum of a flagellar pair. After the next cell division this flagellum resumes growth and grows another 5 um until the middle of the following light period to become the long flagellum of a flagellar pair. Finally after the third cytokinesis these second generation flagella (size class II long flagella) grow another 2-3 µm to reach the mature state during the next light period.

Interestingly it is not only the length of a flagellum that changes during the second cell cycle, but structural and functional changes also occur: since the long and short flagella are associated with different types of microtubular flagellar roots, the basal body of the new long flagellum has to transform its flagellar root system (it grows a second microtubular flagellar root and reduces its 12-stranded microtubular root to a 3-stranded root). In addition the flagellar root system dissociates from the eyespot (the newly formed short flagellum then associates with the eyespot), and the new long flagellum beats with undulatory waves and not with a cilia-type beat pattern as in the previous cell generation. To investigate whether the functional changes in flagellar beat pattern are related to the change in length of the flagellum, we studied flagellar regeneration and behaviour after experimental flagellar amputation (Fig. 8).

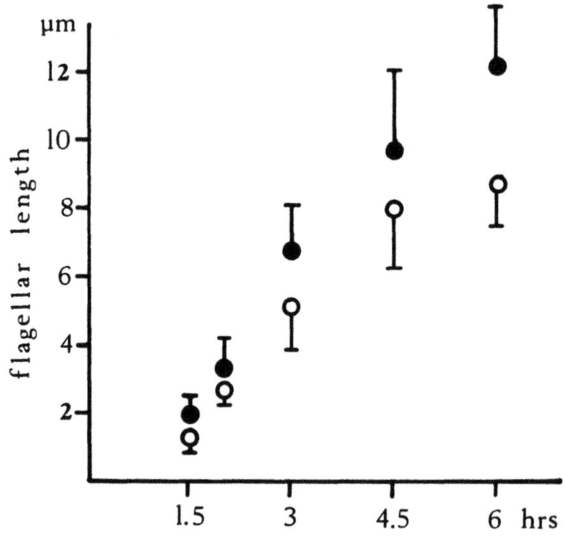

Fig. 8 Kinetics of flagellar regeneration in *Nephroselmis olivacea* in darkness (experimental conditions as in Materials and Methods). Closed circles : long flagellum, open circles: short flagellum. Bars indicate standard deviation for each measurement.

Cells of *N. olivacea* regenerate both flagella after flagellar amputation (Fig. 8). The velocities of flagellar regeneration between the two flagella always differ (Fig. 8). The long flagellum regenerates with an average velocity of 52 nm/min, the short flagellum with about 42 nm/min, i.e. the longer flagellum grows 20 % faster than the short flagellum. To ascertain whether it is a particular basal body that gives rise to the longer or the shorter flagellum during flagellar regeneration, we used the nucleolus as a positional marker, since it is exclusively located in the half of the broad side which is closer to the long flagellum. Without exception the basal body that before deflagellation was associated with the long flagellum also regenerated the longer of the two flagella. Flagellar regeneration within 6-9 hours leads to the establishment of two flagella of unequal length with an L/S-ratio of about 1.4:1. Interestingly, even though the regenerated flagella are relatively short (Fig. 8), cells swim normally i.e. the longer flagellum performs undulating waves and the shorter flagellum beats in a cilia-type fashion. In conclusion the different motile behaviour between the long and the short flagellum is not a consequence of a change in flagellar length which occurs during cell division, but rather represents an independent mechanism that occurs concurrent with the length change and the rearrangements of the basal apparatus during the establishment of size class I long flagella.

DISCUSSION

The results presented in this study demonstrate that N. olivacea attains two dissimilar flagella per cell by extending the development of a single flagellum through more than one cell cycle. Some events associated with cell division apparently activate the further development of a flagellum that was first formed during the previous cell cycle. With respect to the amount of additional flagellar growth the two daughter cells originating from a parental cell are quite dissimilar although they are of equal size: in one daughter cell (receiving the old short flagellum) flagellar growth during the following interphase amounts to about 10 µm (5+5 µm for the two flagella), whereas in the other daughter cell (receiving the old long flagellum) flagellar growth is only 5 µm (for daughter cells with size class III long flagella) or 7 µm (for daughter cells with size class II long flagella). Since N. olivacea is a scaly flagellate the two daughter cells also produce different numbers of flagellar scales. This leads one to speculate that the Golgi apparatus (GA) could be somehow involved in determining the extent of flagellar growth. N. olivacea contains only a single dictyosome and it is possible that the dictyosome divides unequally during cell division and that this results in different scale production rates, different rates of flagellar membrane synthesis and hence different flagellar growth. The change in motile behaviour of flagella and the transformation of the flagellar root system, however, are presumably regulated in other as yet unknown ways.

To what extent can the results obtained with N.olivacea be extended to other algae? We have evidence that the same developmental pattern occurs in another green flagellate, Scourfieldia caeca (unpublished observations). We also believe that the same pattern of flagellar development applies to green algae with flagella of equal length. In Chlamydomonas reinhardtii for example, basal bodies segregate in a semi-conservative fashion, and it is well established that the two flagella are functionally dissimilar (Kamiya et al. 1984). It is therefore likely that a transformation of a basal body from an immature to a mature state occurs in this organism during cell division which somehow affects flagellar function. In Peranema (Euglenophyceae) during cell division the long flagellum shortens (Tamm 1967). We have evidence that the long flagellum of Euglena mutabilis also shortens and transforms to the short flagellum of the next cell generation (Melkonian et al. in preparation). In conclusion it seems possible that flagellar/ basal body development in flagellated algae in general requires more than one cell cycle. This can be regarded as an elegant mechanism to achieve flagellar heterogeneity within a cell. There probably existed a strong selective pressure for the evolution of dissimilar flagella in photoautotrophic flagellates, since two of the most important behavioural responses of flagellates, phototaxis and chemotaxis apparently require functionally different flagella. Our conclusions also seem to be applicable to other developmental processes. The occurrence of two large and complex, but slightly different components in one cell may only be possible through developmental regulation: the development of the two parts of a diatom cell wall for example, occurs in almost the same way as described here for the development of flagella.

ACKNOWLEDGEMENTS

This study was supported by a grant from the Deutsche Forschungsgemeinschaft (Me 658/3-3). M.M. is supported as a Heisenberg fellow by the Deutsche Forschungsgemeinschaft.

REFERENCES

Kamiya,R.,G.B. Witman (1984) J. Cell Biol. 98, 97-107.
McFadden,G.I.,M. Melkonian (1986) Protoplasma 130, 186-198.
Melkonian,M.,H. Robenek (1984) Progr. Phycol. Res. 3, 193-268.
Moestrup,Ø. (1982) Phycologia 21, 427-528.
Moestrup,Ø., H. Ettl (1979) Opera Botanica 49, 3-39.
Pringsheim,E.G. (1946) Pure Cultures of Algae, Cambridge: University Press.
Tamm,S. (1967) J. Exp. Zool. 164, 163-186.

PROTEIN SYNTHESIS IN CHLOROPLASTS

A. Boschetti, E. Breidenbach, S. Leu and J. Clemetson-Nussbaum
Institut für Biochemie, Universität Bern, Freiestraße 3, Ch-3012 Bern

How to study protein synthesis in chloroplasts?

In order to study protein synthesis within chloroplasts it is often necessary to eliminate cytoplasmic protein synthesis. This can be done by two methods:

a) Specific inhibitors of protein synthesis block translation either on chloroplast 70S ribosomes or on cytoplasmic 80S ribosomes. Using e.g. cycoheximide, radioactive amino acids will be exclusively incorporated in proteins synthesized in the organelles. This method enables the _in vivo translation_ of proteins in chloroplasts (and mitochondria) in whole cells or plants to be studied specifically.

b) When pure, intact, photosynthetically active chloroplasts are isolated, cytoplasmic protein synthesis is physically removed. Incorporation of amino acids in such chloroplasts is called _in organello protein synthesis_ . As will be shown later, such chloroplasts also form the starting material for _homologous in vitro translation_ systems.

Procedures for the isolation of chloroplasts are well established for some higher plants and also for some algae, e.g _Euglena_ (Price et al., 1986). In the case of the unicellular green alga _Chlamydomonas reinhardii_ the isolation of intact chloroplasts is rather difficult. However, this organism has so many advantages for studying chloroplast biogenesis - the most important being mutability and genetics - that several laboratories have taken the effort to elaborate protocols for the isolation of chloroplasts from _Chlamydomonas_ . Klein et al. (1983 a,b) and Belknap (1983) used autolysin to weaken the cell wall. They then opened the cells by detergents or gentle mechanical breakage. In our laboratory Dr. Mendiola-Morgenthaler et al. (1985) has developed an isolation procedure starting from synchronously grown mutant cw-15-cells which contain only a very fragile relict of a cell wall. In the upper part of fig. 1 a brief outline of this procedure is given.

Although still contaminated with some cytoplasmic or nuclear fragments, the chloroplasts isolated by this procedure contain almost no cytoplasmic ribosomes or ribosomal subunits. Therefore cytoplasmic mRNA is also believed to be practically absent. The chloroplasts are intact as shown by electron microscopy and by the ferricyanide test, and are photosynthetically active. It is remarkable that O_2 evolution is almost stoichiometrically coupled to CO_2-fixation. These chloroplasts still incorporate (^{35}S)methionine for about 30 min. into proteins. This _in organello_ incorporation is light-dependent,

Algal Development (Molecular and Cellular Aspects)
Ed. W. Wiessner, D.G. Robinson, and R.C. Starr

sensitive to chloramphenicol, but insensitive to cycoheximide (Leu et al. 1984a). The main products are the LS (55 kD) and the 32 kD herbicide-binding membrane proteins (Fig. 4). We are studying protein synthesis with such chloroplast preparations.

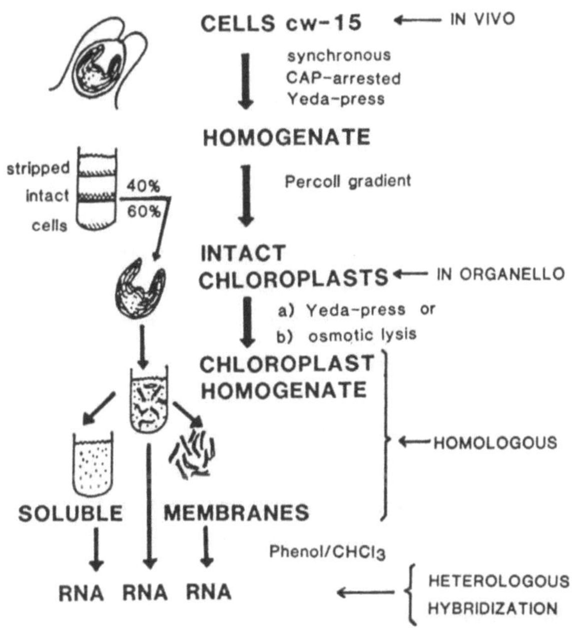

FIGURE 1: Isolation of intact chloroplasts from Chlamydomonas. At the right definition of their lysis and fractionation different translation systems.

TABLE 1: Relationships of the genetic apparatus of chloroplasts to prokaryotes and eukaryotes. 1)Herrmann 1984; 2)Stutz 1984; 3)Bogorad et al.1983; 4)Erickson et al. 1984; 5)Hennig et al. 1986; 6)Berends et al. 1986; 7)Montandon et al. 1984; 8)Whitfeld et al. 1983; 9)Graves et al. 1983; 10)Eberly et al. 1986; 11)Chua et al. 1976; 12)Margulies 1980; 13)Yamamoto et al. 1981; 14)Smith et al. 1978; 15)Adelman et al. 1973; 16)Harrison et al. 1974.

DNA:	Gene clusters(1,2);"-10" and "-35" like sequences(3)	prok.	
	split genes (4)		eukar.
mRNA:	polycistronic pre-mRNA (5)	prok.	
	extensive processing, splicing (6)		eukar.
	no cap, no poly(A); Shine-Dalgarno sequence	prok.	
Ribosomes:	rRNA: homology to E.coli up to 75%	prok.	
	rProteins: homology to E.coli up to 70% (7)	prok.	
tRNA:	3o-31 tRNAs; homology to E.coli often 70-80% (8)	prok.	
	formylated met-tRNA	prok.	
Initiation:	IF-1 43% homologous to E.coli (9)	prok.	
Elongation:	3 elongation factors as in E.coli (10)	prok.	
	Inhibitors	prok.	
Membrane-bound polysomes: (11-16)			eukar.

The protein synthetic apparatus of chloroplasts

As shown in table I most of the components of the protein synthetic apparatus of chloroplasts are more related to prokaryotes than to eukaryotes, which agrees well with the endosymbiontic theory.

However, there are also some pecularities in the genetic apparatus of chloroplasts showing eukaryotic traits. (a) Although polycistronic primary transcripts are often formed from clustered chloroplast genes, there is evidence for an extensive processing of this pre-mRNA in chloroplasts including 5-end shortening and endonucleolytic cleavage to form bi- and monocistronic mature mRNA. (b) Introns have been found in some chloroplast genes. This implies that enzymes for splicing of pre-mRNA must be present in chloroplasts. (c) Polysomes are found on the thylakoid membranes. The binding properties resemble more those realised in the rough endoplasmic reticulum (rER) of eukaryotes than in the binding of polysomes to bacterial membranes. Membrane-bound bacterial ribosomes are immediately released by a simple puromycin treatment. A complete dissociation of thylakoid-bound polysomes, as well as of polysomes of rER from their membranes is achieved only by the combined action of puromycin and high salt. From this observation it is assumed that in chloroplasts and in rER polysomes are attached via not only the nascent polypeptides, but also by a certain affinity of the ribosomes to a putative "receptor" in the membranes.

What is the function of thylakoid-bound polysomes?

In analogy to protein synthesis on rER in eukaryotes it is tempting to assume that thylakoid-bound polysomes preferentially synthesize hydrophobic membrane-proteins which will be immediately inserted into the lipid bilayer. According to this concept, mRNA for membrane proteins should be found predominantly bound to thylakoids, whereas mRNA for soluble proteins should be accumulated on free polysomes of the stromal fraction. To test this hypothesis we have prepared a membrane fraction containing thylakoids and a soluble fraction containing stroma components from chloroplasts (Fig. 1). After washing the membranes with 25 mM Mg-acetate and 200 mM KCl, polysomes remained bound to the thylakoids (Bolli et al. 1981).

From the membrane and soluble fractions the RNA was extracted. The relative mRNA content for a soluble (LS) and a membrane protein (32 kD herbicide-binding) was determined by hybridisation with cloned DNA-fragments containing the genes for the respective proteins (EcoRI fragments 14 and 15 of *Chlamydomonas* chloroplast DNA). Table II shows that the mRNAs for the soluble LS and for the 32-kD membrane-protein were present in the same proportion in RNA extracted either from membranes or from stroma. There is no preferential localization of mRNA for membrane proteins in thylakoid-bound polysomes, nor for mRNA of soluble proteins in free polysomes.

Furthermore, the RNA extracted from membranes contains about twice as much mRNA as that extracted from the soluble part. Together with the results of table III, showing that from membranes about twice as much RNA or ribosomes could be extracted , we have calculated that about 75-80% of all mRNA coding for these two proteins are present in a thylakoid-bound form. The question therefore arises as to the function of the free polysomes.

TABLE II: Relative amounts of mRNA for 32 kD protein and LS in RNA extracted from washed membranes and soluble fraction of isolated, homogenized chloroplasts of *Chlamydomonas reinhardii*. ( ): standard deviation.

RNA from:	Hybridization with EcoRI 14 32 kD	Eco RI 15 LS	32/LS
Chloroplasts	1	1	1
Membranes	1.21 (0.41)	1.12 (0.31)	1.01 (0.19)
Thylakoids	0.73 (0.32)	0.61 (0.23)	1.11 (0.28)

TABLE III: Recovery of ribosomes and RNA from chloroplasts, washed chloroplast membranes and stromal fraction of *Chlamydomonas reinhardii* .

Sample:	Ribosomes:	RNA extracted: $OD_{260} \cdot ml^{-1} \cdot \mu g^{-1}$ chlorophyll	
Chloroplasts	100 %	6.78	100 %
Membranes	55 %	2.95	44 %
Soluble fraction	35 %	1.54	23 %

Evidences for translational regulation in chloroplasts

Obviously, membrane binding of mRNA is not selective. Nevertheless, membranes or other structures may influence protein synthesis in chloroplasts (Minami et al.1984). Indeed, at the present time there are different experimental approaches towards translational regulation in chloroplasts.

In one approach the variations in the rates of protein synthesis during regreening of etiolated plants or algae have been studied (Jenkins et al.1983; Miller et al.1983; Reardon et al.1983; Inanime et al.1985; Fromm et al.1985;

Bouet et al.1985). Several experiments indicate that, in contrast to the nuclear genes coding for chloroplast proteins the chloroplast genes are also transcribed constitutively in the dark. The synthesis of the corresponding proteins, however, can be switched on and off by the light-dark regime.

In our laboratory we have followed an approach involving in vitro translation. In a first experiment we isolated different mRNA-containing fractions from chloroplasts of Chlamydomonas and translated them in the heterologous reticulocyte lysate system (Leu et al.1984). Different translation products were formed when either lysed chloroplasts or thylakoids with bound polysomes, or when phenol-extracted RNA from whole chloroplasts or thylakoids, or when polysomes detached from the membranes by Triton-X-100 were translated. A possible interpretation of these findings is that the translation of chloroplast mRNA is influenced by the presence of some chloroplast components.

A weak point of this experiment was the use of a heterologous translation system. Therefore, we homogenised the Chlamydomonas chloroplasts in the

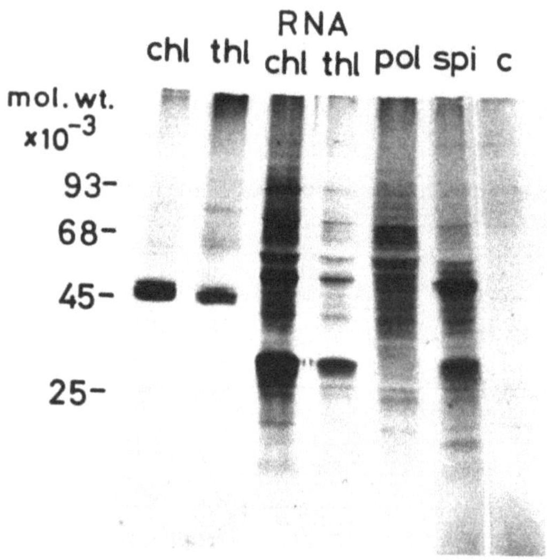

FIGURE 2: Translation products of various mRNA-containing fractions of Chlamydomonas chloroplasts in reticulocyte lysate. chl: lysed chloroplasts; thl: thylakoids; RNA chl, RNA extracted from chloroplasts or thylakoids; pol: polysomes detached from thylakoids; spi: RNA extracted from. spinach chloroplasts; c: blank.

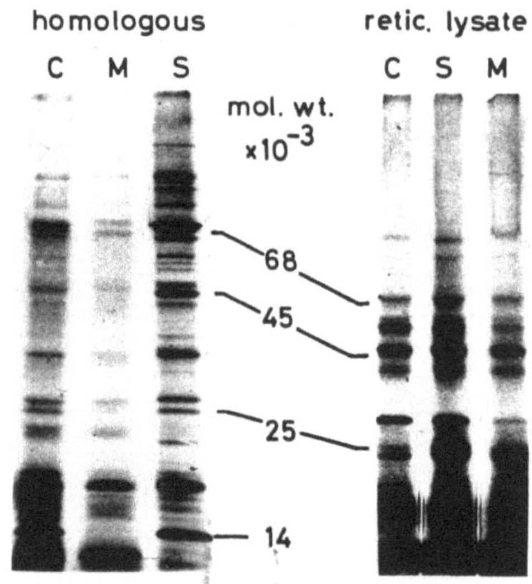

FIGURE 3: Products of homologous translation of chloroplast homogenate (C), membranes (M) and stroma (S) compared to product patterns of heterologous translation of the respectiv extracted RNAs in the reticulocyte lysate.

Yeda-press and isolated the washed membranes and the supernatant stromal fraction (Leu et al.1984b). These fractions were supplemented with an ATP-regenerating system and with (^{35}S-methionine). The resulting homogolous translation patterns are shown in fig. 3. There are qualitative and quantitative differences between the products of stromal and membrane mRNA. However, when from the very same samples the RNA is extracted with phenol and then translated in the reticulocyte lysate, exactly the same patterns were obtained (a result confirming the equal distribution of mRNA between membranes and stroma). This experiment demonstrated that translation of the same set of mRNA is different and influenced by the presence or absence of certain chloroplast components.

Towards a homologous in vitro translation system from chloroplasts

A severe objection against the previous experiment is based on the different product patterns of the homologous and in organello translations. The main products of in organello synthesis, the LS and the 32 kD protein, are almost completely absent from homologous translations. One reason for this could be a mechanism suggested by Mullet et al. (1986) according to which translation might stop or slow down at certain positions on the mRNA especially when translation conditions are not optimal. Recently we have found conditions which have overcome these difficulties and open up future work. Instead of breaking the

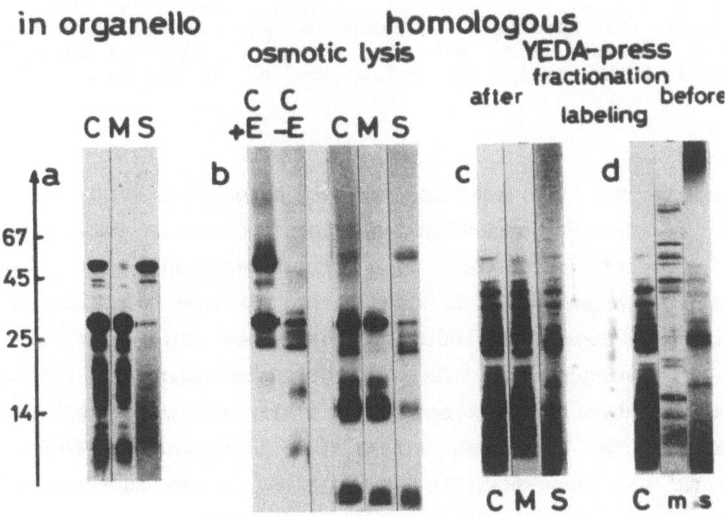

FIGURE 4: Comparison of the product patterns of in organello protein synthesis (a) and of homologous translation of Chlamydomonas chloroplasts homogenized by osmotic shock (b) or Yeda-press (c,d). The chloroplasts or homogenates (C) were separated in membranes and supernatant before (m,s) or after (M,S) incorporation of radioactive (^{35}S)methionine.

chloroplasts by a rather drastic treatment with the Yeda-press we gently lysed them by osmotic shock. When supplemented with an ATP-regenerating system and (^{35}S)methionine a product pattern is obtained similar to that of in organello synthesis (Fig. 4). Apparently the gently homogenization procedure preserves the fidelity of translation. The integrity of some chloroplast structures seems to influence translation.

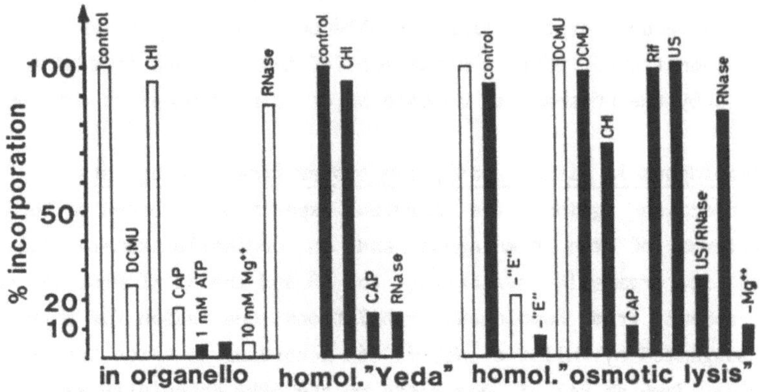

FIGURE 5: Influence of additives on translation in Chlamydomonas chloroplasts in organello and in chloroplast homogenates prepared by Yeda-press or osmotic lysis. Open bars:light; black bars:dark. Standard translation medium for in organello contained no ATP-regenerating system and less than 1 mM Mg^{2+}. Standard medium for homologous translation contained ATP-regenerating system (except when marked -E) and 10 mM Mg^{2+}. CHI: cycloheximide; CAP: chloramphenicol; DCMU: diuron; Rif: rifampicin; US: sonification.

Fig. 5 characterizes the osmotically lysed chloroplast translation system and compares it to the in organello and the Yeda-press homologous system. In organello protein synthesis is light-dependent and inhibited by Mg-concentrations above 1mM. In the dark, ATP does not stimulate methionine incorporation nor does RNase inhibit, showing that chloroplasts are surrounded by an intact envelope. Both the Yeda-press homogenate and the osmotic lysed chloroplasts depend on an ATP-regenerating system and have Mg-optima of about 10 mM. But whereas the Yeda-press translation is susceptible towards RNase, the osmotic lysate is RNase-sensitive only after 5 sec ultrasonication. This suggests that with osmotic lysis the chloroplasts are first opened and then resealed. Only ultrasonication renders the endogenous mRNA attainable to the RNase. We have not yet studied this phenomenon in detail. However, even after sonication the product pattern is still comparable to that on in organello synthesis.

Import of cytoplasmically-made proteins into isolated Chlamydomonas chloroplasts

Not all chloroplast proteins are made within the chloroplast. Many enzymes and submembrane complexes of photosynthetically active chloroplasts are composed stoichiometrically of polypeptides synthesized in the chloroplast, and of those which are coded in the nucleus, synthesized as precursors in the cytoplasm and imported and processed in the chloroplast. To understand chloroplast biogenesis, studies on protein synthesis in the chloroplast have to be parallelled by those on protein import. In vitro import of cytoplasmically-made protein precursors have been studied with chloroplasts of different species (Ellis 1981; Grossman et al.1982; Kloppstech et al.1984; Mishkind et al.1985; Robinson et al.1985; Vierling et al.1986).

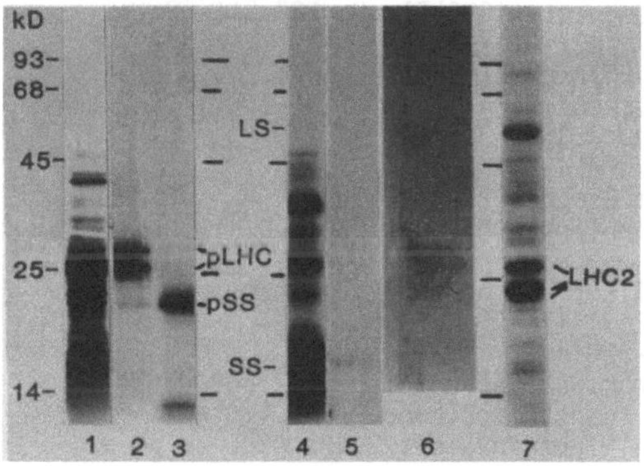

FIGURE 6: Import of precursors of LHC II-proteins and SS into isolated Chlamydomonas chloroplasts. (1) Poly (A)-RNA from Chlamydomonas translated in a wheat germ extract; (2) precursors of LHC II-proteins and of SS(3) immunoprecipitated from 1; (4)radioactive products incorporated in reisolated chloroplasts; (5) autoradiogram of isolated carboxylase-holoenzyme and (6) of crossed immunoelectroferogram of membranes reacted with antiserum against 25 kD LHC II-protein;(7) same as 4, but Coomassie stained.

In Fig. 6 we show that Chlamydomonas chloroplasts, isolated by our procedure, are also capable of importing invitro precursors of the small subunit of ribulose-biphosphate carboxylase (SS) and of the apoproteins of the light-harvesting complex II (LHC II). According to Bartlett et al.(1982), poly(A)-containing RNA from Chlamydomonas has been translated in a wheat germ extract. Among the products, precursors of LHC II-proteins and SS were found by immunoprecipitation. The supernatant of the translation mixture was incubated with intact chloroplasts from Chlamydomonas . After protease treatment, intact chloroplasts were reisolated by centrifugation through Percoll. The chloroplasts were then separated in supernatant and membranes. In the carboxylase-holoenzyme, isolated by centrifugation from the supernatant, radioactively labelled SS was found. In the membranes radioactive LHC II-proteins could be identified by crossed immunoelectrophoresis.

With the development of procedures for the isolation of intact chloroplasts from Chlamydomonas , which are capable of performing protein synthesis and protein import in vitro, this organism will become an even more attractive organism for studying biogenesis on chloroplasts.

Acknoledgements : The cloned DNA probes were a gift of Dr. D. Herrin Tampa/Athens, USA. We greatfully acknowledge the financial support of the Swiss National Foundation for Scientific Research and of the Roche Research Foundation.

References

Adelman,M.R.,D.D.Sabatini,G.Blobel(1973)J.Cell Biol.56,206-229.
Bartlett,S.G.,A.R.Grossman,N.H.Chua(1982)In:Meth.chloroplast molec.biology,eds.M.Edelman et al.Elsevier Biomed.Press,Amsterdam,p.p.1081-1091.
Belknap,W.R.(1983)Plant.Physiol.72,1130-1132
Berendo,T.,Q.Kubicek,J.E.Mullet(1986)Plant Molec.Biol.6,126-134.
Bogorad,L.,E.J.Gubbius,S.O.Jolly,E.T.Krebbers,I.M.Larrinua,K.M.T.Muskavitch,S.R.Rodermel,A.Subramanian,A.Steinmetz(1983)In:Gene Structure and Regulation in Development. A.R.Liss,Inc.New York, pp.13-31.
Bolli,R.,L.Mendiola-Morgenthaler,A.Boschetti(1981)Biochim.Biophys.A.653,276-287.
Bonet,C.,R.Schantz,G.Dubertret,B.Pineau,G.Ledoigt (1986)Planta 167,511-52o.
Chua,N.H.,G.Blobel,P.Siekevitz,G.E.Palade,J.Cell Biol.71,497-514.
Eberly,S.L.,G.H.Spremulli,L.L.Spremulli (1986)Arch.Biochem.Biophys.245,338-347.
Ellis,R.J.(1981)Ann.Rev.Plant Physiol.32,111-137.
Erickson,J.M.,M.Rahire,J.D.Rochaix(1984)EMBO J.3,2753-2762.
Fromm,H.,M.Devic,R.Fluhr,M.Edelman(1985)EMBO J.4,291-295.
Graves,M.C.,L.L.Spremulli(1983)Arch.Biochem.Biophys.222,192-199.
Grossman,A.R.,S.G.Bartlett,G.W.Schmidt,J.E.Mullet,N.H.Chua(1982)J.Biol.Chem.257, 1558-1563.
Harrison,T.M.,G.G.Brownlee,C.Milestein(1974)Eur.J.Biochem.47,613-62o.
Hennig,J.,R.G.Herrmann(1986)Mol.Gen.Genet.2o3,117-128.
Herrmann,R.,(1984)Ber.Dtsch.Bot.Ges.97,335-350.
Inamine,G.,B.Nash,H.Weissbach,N.Brot(1985)Proc.Natl.Acad.Sci.U.S.82,5690-5694.
Jenkins,G.,M.R.Hartley,J.Bennett(1983)Phil.Trans.R.Soc.Lond.B 303,419-431.
Klein,U.,C.Chen,M.Gibbs,A.Platt-Aloia(1983a)Plant Physiol.72,481-487.
Klein,U.,C.Chen,M.Gibbs(1983b)Plant Physiol.72,488-491.
Kloppstech,K.,G.Meyer,G.Schuster,I.Ohad(1985)EMBO J.4,19o1-1909.
Koharn,B.D.,E.Harel,P.R.Chitnis,J.P.Thornber,E.M.Tobin(1986)J.Cell Biol.102,972-981.
Leu,S.,L.Mendiola-Morgenthaler,A.Boschett(1984a)Febs Lett.166,23-27.
Leu,S.,R.Bolli,L.Mendiola-Morgenthaler,A.Boschetti(1984b)Planta 160,204-211.
Margulies,M.M.,(198o)Biochim.Biophys.A.606,13-19.
Mendiola-Morgenthaler,L.,S.Leu,A.Boschetti(1985)Plant Sci.38,33-39.
Miller,E.,J.E.Jugenson,E.M.Reardon,C.A.Price(1983)J.Biol.chem.258,14478-14484.
Minami,E.,A.Watanabe(1984)Arch.Biochem.Biophys.235,562-570.
Mishkind,M.L.,S.R.Wessler,G.W.Schmidt(1985)J.Cell.Biol.100,226-234.
Montandon,P.E.,E.Stutz(1984)Nucl.Acids Res.12,2851-2859.
Mullet,J.E.,R.R.Klein,A.R.Grossman(1986)Eur.J.Biochem.155,331-338.
Reardon,E.M.,C.A.Price(1983)Arch.Biochem.Biophys.226,433-440.
Robinson,C.,R.J.Ellis(1985)Eur.J.Biochem.152,67-73.
Smith,W.P.,P.C.Tai,B.D.Davis(1978)Proc.Natl.Acad.Sci.U.S.75,814-817.
Stutz,E.,(1984)Bot.Helv.94,145-159.
Vierling,E.,M.L.Mishkind,G.W.Schmidt,J.L.Key(1986)Proc.Natl.Acad.Sci.U.S.83,361-365.
Whitfeld,P.R.,G.Zurawski,W.Bottomley(1983)In:Manipulation and expression of genes in eukaryotes.Acad.Press.Australia,pp.247-254.
Yamamoto,T.,J.Burke,G.Autz,A.T.Jagendorf(1981)Plant Physiol.67,940-949.

COMPARISON OF ENVELOPE MEMBRANES FROM HIGHER PLANTS AND ALGAE PLASTIDS AND OF OUTER MEMBRANES FROM CYANOBACTERIA (BLUE-GREEN ALGAE)

J. Joyard, M.A. Block, A.-J. Dorne and R. Douce

Laboratoire de Physiologie Cellulaire Végétale, D.R.F., U.A. CNRS n° 576, CENG and USTMG, 85 X, F-38041 Grenoble-cédex, France.

The plastid envelope is a fascinating membrane system. In higher plants and in several algae, the envelope consists of a pair of membranes although it is a more complex structure in numerous algae such as Euglena (Lefort-Tran, 1981). Envelope membranes provide a flexible boundary between the plastid and the surrounding cytosol (Douce et al, 1984). Therefore, a major function of envelope membranes is the selective control of metabolite transport, owing to specific translocators located on the inner membrane (Heldt and Heber, 1981). Besides this role in photosynthesis regulation, the envelope membranes are involved in plastid biogenesis. They regulate the uptake and transport of plastid proteins that are synthesized on cytoribosomes (Chua and Schmidt, 1979). In addition, they can catalyze the synthesis of characteristic plastid glycerolipid and prenylquinones (Douce et al, 1984); these studies were made possible owing to the development of efficient methods to purify large amounts of envelope membranes from higher plant chloroplasts. With the exception of Euglena, few studies were done with algae plastid envelopes (Blée, 1981; Brandt, 1982/83), mostly due to the difficulty to prepare intact chloroplasts. However, this has recently been achieved for Chlamydomonas (Mendiola-Morgenthaler et al, 1985) and it is therefore possible to compare some properties (lipid content and role in glycerolipid biosynthesis) of envelope membranes from higher plants and algae; this is the purpose of this article. In addition, further comparison with the outer membranes of cyanobacteria is most interesting since it is generally assumed, without experimental evidences, that the inner envelope membrane from chloroplasts could derive from the plasma membrane of the prokaryotic ancestor whereas the outer envelope membrane of chloroplasts could derive from the endomembrane system of the protoeukaryote that engulfed the free living prokaryote.

POLAR LIPID COMPOSITION OF PLASTID ENVELOPE MEMBRANES

Treatment of isolated chloroplasts with a non-penetrant protease, such as thermolysin, prior to envelope purification and lipid determination is necessary, at least in higher plants, to destroy a galactolipid :galactolipid galactosyltransferase (Dorne et al, 1982) which catalyzes the exchange of galactoses

Algal Development (Molecular and Cellular Aspects)
Ed. W. Wiessner, D.G. Robinson, and R.C. Starr

between galactolipids (Van Besouw and Wintermans, 1978) and led to the modification of the true polar lipid composition of envelope membranes: MGDG (for abbreviations, see Table 1) being destroyed to form diacylglycerol and unnatural galactolipids (TGDG and TTGDG). This enzyme is localized on the cytosolic side of outer envelope from spinach chloroplasts (Dorne et al, 1982; Heemskerk et al, 1986) and from pea etioplasts (Soll et al, unpublished data). It is not known whether the galactolipid:galactolipid galactosyltransferase occurs in algae chloroplasts, but if it does, it is probably to a low extent since only very low levels of diacylglycerol (17 µg/mg proteins) were detected in envelopes from Euglena chloroplasts (Blée, 1981) whereas the total lipid composition of Chlamydomonas envelope shows no TGDG or TTGDG (Mendiola-Morgenthaler et al, 1985) which are witnesses of the galactosyltransferase activity.

Plastid envelope membranes are characterized by a high glycolipid and a low phospholipid content (Table 1) (Douce et al, 1973): as all plastid membranes, envelopes are devoid of PE -the major polar lipid in extraplastidial membranes and of cardiolipin -a specific inner mitochondrial membrane lipid. Except for PC (see below), table 1 shows the similarity in the lipid composition of plastid membranes from higher plants and algae, and membranes from cyanobacteria or Prochloron, a prokaryote which belongs to a special algae division, the prochlorophyta (Lewin, 1981).

In higher plant chloroplasts, the polar lipids of both envelope membranes are identical, but the proportions in which they are present are different (Cline et al, 1981; Block et al, 1983a). In outer envelope membranes from thermolysin-treated spinach chloroplasts, MGDG and DGDG are present in the ratio 0.6:1 whereas PC and PG are in the ratio 3:1. The situation is reversed in the inner membrane, which is highly comparable to thylakoids. In this case, MGDG and DGDG are present in a ratio of 1.6:1, and PC and PG in a ratio of 0.5-0.8:1, respectively (Table 1). Although the separation of outer and inner membranes from algae chloroplasts has not yet been achieved, the results with total envelopes and thylakoids from Chlamydomonas (Mendiola-Morgenthaler et al, 1985) are in agreement with the polar lipid distribution in higher plants chloroplasts. Interestingly, in the cyanobacteria Anacystis nidulans, thylakoids and plasma membrane (inner membrane) have also a similar polar lipid composition (Omata and Murata, 1983) (Table 1). The higher level of DGDG in the cell envelope of Anacystis nidulans (Murata et al, 1981), when compared with thylakoids, is similar to the difference observed between envelope membrane and thylakoids from higher plant or algae chloroplasts and reflects the lipid composition of the outer membrane, which has a higher DGDG content than the other plastid membranes (Table 1). A difference is sometimes detected among glycolipids: minor amounts of monoglucosyldiacylglycerol (MgluDG) are present in some cyanobacteria (Table 1). As we shall see later, in prokaryotes, this glycolipid is a precursor of MGDG synthesis and can accumulate, although to a small extent.

The major difference between chloroplasts and cyanobacteria is related to PC: prokaryotes are usually devoid of PC whereas in higher plant envelopes, PC represents about 20 % of the polar lipid content. In some eukaryotic algae, such as Chlamydomonas, PC is replaced by an unusual lipid: diacylglyceryl-O-(N,N,N-trimethyl) homoserine (DGTS). This betaine-containing lipid is somewhat stucturally related to PC and is widely distributed among green algae, pteridophytes and bryophytes (Eichenberger, 1982; Sato and Furuya, 1984). The

Table I - **Glycerolipid composition of membranes from prokaryotes and organelles from algae and higher plants**

	MgluDG	MGDG	DGDG	SQDG	PC	DGTS	PG	PI	PE	DPG
PROKARYOTES										
- **Prochloron** sp.[a]	3.2	55.3	10.4	25.7	0	0	5.4	0	0	0
- **Anabaena variabilis**[b]	1	54	17	11	0	0	17	0	0	0
- **Anacystis nidulans**										
total[c]	0	50.8	19	15.9	0	0	14.3	0	0	0
thylakoids[d]	0	55	12	7	0	0	22	0	0	0
cytoplasmic membranes[d]	0	63	12	7	0	0	18	0	0	0
cell envelope[e]	0	53	24	5	0	0	18	0	0	0
ALGAE PLASTIDS										
- **Chlamydomonas reinhardii**[f]										
thylakoids	-	55	20	13	0	5	6	2.7	0.6	0
envelope	-	27	31	8	0	21	3.5	4.1	4.6	0
HIGHER PLANTS PLASTIDS										
- **Spinach** chloroplasts[g]										
thylakoids	-	52	26	6.5	4.5	-	9.5	1.5	0	0
inner envelope membrane	-	49	30	5	6	-	8	1	0	0
outer envelope membrane	-	17	29	6	32	-	10	5	0	0
total envelope	-	32	30	6	20	-	9	4	0	0
- **Pea** etioplasts[h]										
prothylakoids	-	42	35	6	9	-	5	2	0	0
envelope	-	34	31	6	17	-	5	4	0	0
- **Cauliflower** buds non-green plastids[i]	-	30	29	6	19	-	8	5	2	0.6
MITOCHONDRIA										
- **Mung bean** hypocotyls[j]										
total	-	0	0	0	36	-	1	2.5	46	14
inner membrane	-	0	0	0	29	-	1	2	50	17
outer membrane	-	0	0	0	68	-	2	5	24	0

References: a, Murata and Sato (1982); b, Murata and Sato (1983); c, Allen et al (1966); d, Omata and Murata (1983); e, Murata et al (1981); f, Mendiola-Morgenthaler et al (1985); g, Block et al (1983a); h, Soll et al (unpublished data); i, Journet and Douce (1985); j, Bligny and Douce (1980); -, not analyzed. Abbreviations: MgluDG, monoglucosyldiacylglycerol; MGDG, monogalactosyldiacylglycerol; DGDG, digalactosyldiacylglycerol; SQDG, sulfolipid; PC, phosphatidylcholine; PG, phosphatidylglycerol; PI, phosphatidylinositol; PE, phosphatidylethanolamine; DPG, cardiolipin.

presence of DGTS, instead of PC and in the same proportions, in envelope membranes from *Chlamydomonas* is most interesting but it is not known whether this can be a general feature in all DGTS-containing plants.

In chloroplasts, PC is localized in the outer envelope membrane (Cline et al, 1981; Block et al, 1983a). It is not known whether in DGTS-containing envelope the same localization in outer envelope occurs, but this is highly likely. This situation, together with the presence of sterols in envelope membranes (see Douce et al, 1984), could reflect a possible dual origin of the envelope membrane. However, if the outer envelope is shown as homologous with the endomembrane system of the protoeukaryotic host cell, it is difficult to understand why the outer envelope membrane contains large amounts of typical cyanobacteria lipids, such as galactolipids and sulfolipid, and does not contains PE, a characteristic endomembrane polar lipid. Since PC, in the outer chloroplast envelope, is solely localized in the outer leaflet (Dorne et al, 1984), it could be possible that at least only the leaflet facing the cytosol could derive from the endomembrane system. We believe that this hypothesis is unlikely for the following reasons: (a) cytochemical analyses have shown that both envelope membranes, together with the internal thylakoids membrane system, share characteristic chemical constituents which are responsible for the specific staining of plastid membranes within the plant cell (see Douce et al, 1984); (b) galactolipids and sulfolipid are accessible to specific antibodies from the cytosolic side of isolated intact chloroplasts (see Douce et al, 1984). Therefore, the major conclusion for these observations is that both the outer and the inner envelope membrane have a typical plastid origin. However, the problem of the origin of PC (or DGTS) in the outer envelope membrane remains to be elucidated. Since envelope membranes are probably unable to synthesize PC (Joyard and Douce, 1976), it is possible that PC accumulation in the outer leaflet of the outer envelope could be due to PC transfer from endoplasmic reticulum to the envelope, for instance via a phospholipid transfer protein (Miquel et al, 1982). This is probably the same problem for DGTS-containing envelope membranes, since DGTS is typically a eukaryotic lipid (Eichenberger, 1982). Therefore, one can suggest that during the course of the integration of the prokaryote within the protoeukaryotic ancestor, one of the possible adaptation was addition of PC (or DGTS) into the outer leaflet of the outer membrane. Indeed cyanobacteria have a highly asymmetric outer membrane (Peschek, 1984) which contains much less polar lipids in the outer leaflet than in the inner one. Such a situation would highly favor integration of additional lipid molecules in the outer leaflet. Interestingly, the outer membrane of other cellular organelles, such as mitochondria (Bligny and Douce, 1980), is also strongly enriched in PC.

Another characteristic feature of polar lipids is their fatty acid content. Indeed, the galactolipids, sulfolipid, phospholipids from envelope and thylakoid membranes of spinach chloroplasts have been analyzed with respect to proportions, positional distribution, and pairing of fatty acids (Siebertz et al, 1979). All the specific characteristics of the diacylglycerol moiety of lipids known from previous analyses of thylakoid lipids or whole leaves (Heinz, 1977) were also found in envelope lipids. For example, C16 fatty acids, whenever analyzed in typical chloroplast lipids are mainly concentrated at the sn-2 position of the glycerol backbone. In contrast, PC is separated from the aforementionned lipids by the fact that it excludes C16 fatty acids (in this case 16:0) from its sn-2 position and therefore may be considered to belong to a different group. The structure of algae lipids is very similar to those of higher plants at least when one consider that fatty acid distribution is determined by chain length rather than unsaturation (Harwood and Russell, 1984). In this context, it is interesting to note that prokaryotic cyanobacteria direct C16 fatty acids to sn-2 position and C18 fatty acids to sn-1 position of the glycerol backbone (Heinz, 1977; Murata and Sato, 1982). Analyses of the diacylglycerol part of glycolipids led to the conclusion that these structures can be separated into two groups: the first one has C16 or C18 fatty acids at sn-1 position of the glycerol backbone and C16 fatty acids at the sn-2 position whereas the second one has C16 or C18 fatty acids at sn-1 position of the glycerol backbone and C18 fatty at sn-2 positions. Since the first structure is characteristic for cyanobacteria therefore it is called "prokaryotic" structure, whereas the second one is called "eukaryotic" structure and is present in all eukaryotic cells (Heinz, 1977). Galactolipids, and especially MGDG, can have both prokaryotic and eukaryotic structure: in plants such as pea (18:3 plant), only the eukaryotic structure is present in MGDG whereas in plants such as Anthriscus (16:3 plant) only the prokaryotic structure is present in MGDG, spinach (another 16:3 plant) contains both structures (Heinz, 1977). In MGDG from algae, such as Chlorella, Ulva or Enteromorpha C16 polyenoic fatty acids (16:2 or 16:4, depending of algae species) are also localized at sn-2 position (Heinz, 1977) thus suggesting that these algae behave like "16:3 plants". As we shall see below, these different structures for MGDG could reflect their biosynthetic routes.

Finally, the major fatty acids present in higher plant galactolipids are polyunsaturated fatty acids (16:3 and/or 18:3). In algae, the situation is more complicated since fatty acids with 4 double-bonds and very long chain fatty acids also occur in MGDG, DGDG or SQDG, especially in brown algae (Harwood and Russell, 1984). This is in contrast with the situation in most cyanobacteria, such as Anacystis nidulans (Nichols et al, 1965; Omata and Sato, 1983), or in Prochloron (Murata and Sato, 1983) which contain saturated and monounsaturated fatty acids (see Peschek, 1984). However, this is not a general situation since Anabaena variabilis contains polyunsaturated fatty acids in appreciable amounts (Nichols et al, 1965). Therefore, the degree of unsaturation cannot be entirely considered as a characteristic feature of eukaryotic structures.

THE ROLE OF PLASTID ENVELOPE MEMBRANES IN GLYCEROLIPID BIOSYNTHESIS

Glycerolipid biosynthesis requires the assembly of three parts: fatty acids, glycerol and a polar head (galactose, for galactolipids; sulfoquinovose, for SQDG; phosphorylglycerol for PG).

In plant cells, plastids are the sole source of fatty acids (16:0, 18:0 and 18:1) for membranes of developing organelles and serve as precursors of glycolipids and phospholipids (see Harwood and Russell, 1984; Stumpf, 1980; Roughan and Slack, 1982). It is not yet known whether the same is true for eukaryotic algae. During their de novo biosynthesis, their elongation and desaturation to 18:1, the plastid fatty acids remain attached to ACP (acyl carrier protein), therefore, the major products of fatty acid biosynthesis in the plastid stroma are 16:0-ACP and 18:1-ACP. These molecules are destined for further metabolism outside the chloroplast (for biosynthesis of phospholipids such as PC and PE) or within the organelle (for galactolipid, SQDG and PG biosynthesis).

The first step in glycerolipid biosynthesis is the acylation of a water soluble precursor, sn-glycerol 3-phosphate, which is provided by the cytosol. Douce and Guillot-Salomon (1970) demonstrated that isolated chloroplasts are able to catalyze the acylation of sn-glycerol 3-phosphate to form phosphatidic acid, and then by dephosphorylation, to form diacylglycerol, the precursor of galactolipids. In addition to higher plant chloroplasts, all plastids analyzed so far -proplastids (Journet and Douce, 1985), amyloplasts (Fishwick and Wright, 1980), chromoplasts (Liedvogel and Kleinig, 1979) as well as chloroplasts from algae (Boehler and Ernst-Fonberg, 1976; Chammai and Schantz, 1979; Jelsema et al, 1982) are able to acylate sn-glycerol 3-phosphate for further metabolism into specific glycerolipids. In higher plant plastids, all the enzymes involved in this synthesis, known as the Kornberg-Pricer pathway, have been localized on envelope membranes (Joyard and Douce, 1977) and more precisely on the inner envelope membrane (Block et al, 1983a,b). The whole pathway is summarized in Fig. 1A. The first acyltransferase is "soluble", but probably functions in close association with the envelope. It catalyzes the acylation of sn-1 position of sn-glycerol 3-phosphate to yield lysophosphatidic acid. The second acyltransferase is firmly membrane bound and directs C16 fatty acids to sn-2 position of the glycerol (Joyard et al, 1979; Frentzen et al, 1983). Finally diacylglycerol is then formed by a specific phosphatidic acid phosphatase (Joyard and Douce, 1979) localized on the inner membrane (Block et al, 1983b).

All together, these results demonstrate that owing to their envelope membranes, the chloroplasts -and all kind of plastids- are an important site of phosphatidic acid and diacylglycerol within the plant cell (see Fig. 1). However, it is not yet known how diacylglycerol molecules having C18 fatty acids at both position (i.e. having the eukaryotic structure) are synthesized, but it has been proposed that they originate from PC formed outside plastids (Roughan and Slack, 1982; Heinz and Roughan, 1983).

A question which also remains to be elucidated is whether envelope membranes are the sole site of phosphatidic acid synthesis in chloroplasts from all plant species. For instance, it is not known whether growing thylakoids from developing plastids contain, or are devoid of, the enzymes of the Kornberg-Pricer pathway. In fact, proper characterization of a phosphatidic acid and diacylglycerol biosynthesis within thylakoids is hampered by several limitations. First, thylakoids are almost never analyzed with respect to contamination by envelope membranes or extraplastidial membranes containing the enzymes of the Kornberg-Pricer pathway. For instance, we have found that almost half of the estimated amount of envelope membranes from spinach chloroplasts was recovered in the thylakoid fraction when intact chloroplasts were fractionated after an osmotic shock. Furthermore, the inner envelope membrane and thylakoids have almost the same polar lipid composition (see above), a situation which highly favor artifactual fusions, especially after osmotic shock (Lucy and Ahkong, 1986), and therefore may lead to wrong observations for enzymes distribution. In addition, thylakoid and inner envelope membrane have very tight relationship during the first stages of plastid development, therefore these two membrane systems may show the same enzymatic equipment necessary for plastid biogenesis. Furthermore, the stage of development of the plant analyzed is obviously most important (Brangeon and Forchioni, 1986). Finally, some species specificities cannot be ruled out, and the results obtained showing acylation of _sn_-glycerol 3-phosphate in growing _Chlamydomonas_ thylakoids by biochemical (Jelsema et al, 1982) or cytochemical (Michaels et al, 1983) investigations cannot be extended to mature spinach thylakoids without caution.

Douce (1974) has demonstrated that the chloroplast envelope is also the site of UDP-galactose (UDP-gal) incorporation into galactolipids in leaf cells and that the activity is not associated with endoplasmic reticulum. This ability was confirmed on other plastid types (see Douce and Joyard, 1980). At least two distinct enzymes are responsible for the synthesis of MGDG and DGDG. One is specific for the synthesis of MGDG, using UDP-gal as a galactosyl donor (Fig. 1B). The second enzyme is responsible for the formation of DGDG and is either a UDP-gal:MGDG galactosyltransferase (not yet characterized) or a galactolipid:galactolipid galactosyltransferase (Van Besouw and Wintermans, 1978; Heemskerk, 1986). When UDP-gal is supplied to isolated envelope membranes from higher plant chloroplasts, MGDG formation is readily followed by appearance of DGDG, TGDG, TTGDG or even higher homologues, which is in agreement with the functionning of both enzymes whereas in envelope membranes isolated from algae, such as _Euglena_ (Blée, 1981) or _Chlamydomonas_ (Mendiola-Morgenthaler et al, 1985), DGDG is the

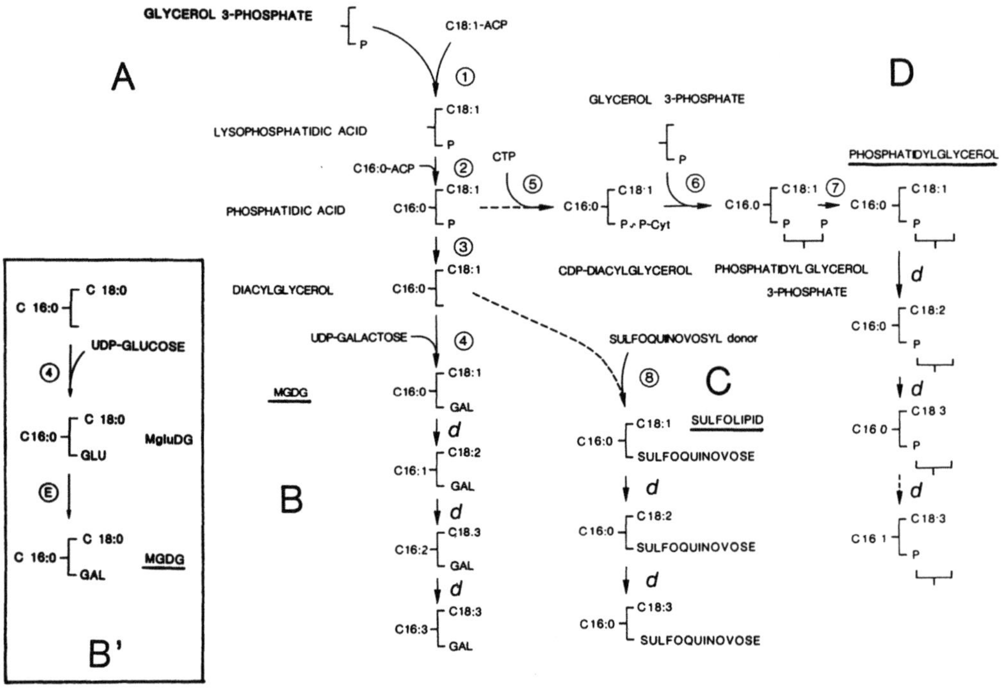

FIGURE 1 - **Pathways of glycerolipid biosynthesis in chloroplasts and cyanobacteria**. Isolated intact chloroplasts and cyanobacteria are able to synthesize MGDG, SQDG and PG. The backbone for these polar lipids comes from phosphatidic acid and diacylglycerol, which are synthesized on the chloroplast envelope (**A**) by the Kornberg-Pricer pathway, owing to an acyl-ACP:sn-glycerol 3-phosphate acyltransferase (1), which is responsible for the synthesis of lysophosphatidic acid; to an acyl-ACP:monoacylglycerol 3-phosphate acyltransferase (2), which is responsible for the formation of phosphatidic acid; and to a specific phosphatidic acid phosphatase (3), which yields diacylglycerol. MGDG is then synthesized (**B**) by galactosylation of diacylglycerol on the inner envelope membrane (4). In cyanobacteria (**B'**), MGDG is synthesized by epimerization (**E**) of MgluDG (see text). All these steps have been demonstrated on isolated envelope membranes. PG synthesis (**D**) involves the formation of intermediates such as CDP-diacylglycerol (5) and phosphatidylglycerolphosphate (6) which is dephosphorylated into PG (7). Sulfolipid biosynthesis (**C**) occurs also in isolated chloroplasts and cyanobacteria. In chloroplasts and in the few cyanobacteria containing polyunsaturated fatty acids, desaturation (d) occurs after the formation of polar lipids, on MGDG, SQDG and on PG. The molecular species synthesized have C18 fatty acids in the sn-1 position of the glycerol backbone and C16 fatty acids in the sn-2 position. There is still controversial information about the synthesis of polar lipids with C18/C18 backbone and also about DGDG synthesis.

main galactolipid to be synthesized. These experiments gives apparent support for the direct galactosylation of MGDG to form DGDG since algae envelope lack diacylglycerol for an active MGDG synthesis (Blée, 1981).

Galactolipid formation has also been studied in some cyanobacteria, such as Anabaena variabilis (Sato and Murata, 1982a,b,c), and in Prochloron (Sato and Murata, personal communication). Most of the pathway is very similar to that found in chloroplasts, however, the last steps involve the transfer of a glucosyl group, instead of a galactosyl group, from UDP-glucose to diacylglycerol (Fig. 1B'). The product formed, MgluDG, is found only in minute amounts in the membranes (see Table 1) since it is rapidly transformed into MGDG by epimerisation of the glucosyl group (Sato and Murata, 1982c) (Fig. 1B'). Chloroplasts apparently lack this epimerase activity but another type of epimerase is present in the cytosol which transform UDP-glucose into UDP-gal, therefore the source of UDP-gal in higher plant cells is the cytosol.

All together, these results demonstrate that owing to their envelope membranes, chloroplasts are able to synthesize their major lipid constituents, galactolipids, and that the pathway for this synthesis as well as the diacylglycerol structure was highly similar to that which is actually found in free-living or symbiotic prokaryotes (Fig. 1). This is also valid for other plastid polar lipids such as SQDG or PG (Fig. 1C,D). For instance, chloroplasts incubated in presence of $^{35}SO_4^{2-}$ (Kleppinger-Sparace et al, 1985; Joyard et al, 1986) or [1-^{14}C]-acetate (Joyard et al, 1986) are able to synthesize SQDG. Joyard et al (1986) have demonstrated that newly synthesized molecules were formed from diacylglycerol having the "prokaryotic" structure. The same is true for PG. Chammai and Schantz (1979) have demonstrated that Euglena chloroplasts contain the complete array of PG biosynthetic machinery. A de novo PG biosynthesis was also demonstrated in spinach chloroplasts (Mudd and de Zacks, 1981; Sparace and Mudd, 1982) and all the enzymes involved were localized in inner envelope membranes (Andrews and Mudd, 1985) (Fig. 1D). Interestingly, PG and SQDG are also synthesized in Anabaena variabilis (Murata and Sato, 1982). Therefore, higher plant or algae chloroplasts behave like cyanobacteria for the synthesis of glycerolipids having the prokaryotic structure, but an additional pathway is probably present in higher plants for the formation of eukaryotic glycerolipids.

CONCLUSION

This short review was restricted to the glycerolipid content and biosynthesis in plastid membranes, and especially envelope membranes. However, numerous additional properties support the hypothesis that envelope membranes are a permanent structure among the plant kingdom. Indeed, the composition of the lipidic phase of envelope membrane is highly conserved in higher plants and algae. For instance, Chlamydomonas envelope membranes show the same absorption spectrum (Mendiola-Morgenthaler et al, 1985) as envelope from spinach chloroplasts or potato tuber amyloplasts (Douce et al, 1984), thus demonstrating the presence of carotenoids in all these membranes. In addition, cyanobacteria outer membranes are also deep yellow. Furthermore, the enzymes involved in the biogenesis are present in envelope from higher plant or algae chloroplasts as well as in the outer membranes from cyanobacteria.

However, an essential prerequisite for growth and differentiation of higher plants or algae is the ability of plastids to meet the changing requirements of the cell. It is clear therefore that, in higher plants as well as in algae, many developmental transitions and species specificities are associated with, or dependant upon, marked changes in the specific equipement of plastid envelope membranes.

ACKNOWLEDGEMENTS

The help of Mrs. M.-H. Bory in preparing the manuscript is gratefully acknowledged.

REFERENCES

Allen, C.F., Hirayama, O. and Good, P., (1966) in: Biochemistry of Chloroplasts, volume 1, T.W. Goodwin, ed., pp. 195-200, Academic Press, London.
Andrews, J. and Mudd, J.B. (1985) Plant Physiol. 79, 259-265.
Blée, E. (1981) Thèse de Doctorat d'Etat. Université de Strasbourg, France.
Bligny, R. and Douce, R. (1980) Biochim. Biophys. Acta 617, 254-263.
Block, M.A., Dorne, A.-J., Joyard, J. and Douce, R. (1983a) J. Biol. Chem. 258, 13281-13286.
Block, M.A., Dorne, A.-J., Joyard, J. and Douce, R. (1983b) FEBS lett. 169, 111-115.
Boehler, B.A. and Ernst-Fonberg, M.L. (1976) Arch. Biochem. Biophys. 175, 229-235.
Brandt, P. (1982/83) Plant Sci. Lett. 28, 237-244.
Brangeon, J. and Forchioni, A. (1986) in: Regulation of chloroplast differentiation, G.Akoyunoglou and H. Senger, eds., pp. 141-146. Alan R. Liss, New-York.
Chammai, A. and Schantz, R. (1979) in: Advances in the Biochemistry and Physiology of Plant lipids, L.-A Appelqvist and C. Liljenberg, eds., pp.381-386, Elsevier, Amsterdam.
Chua, N.-H. and Schmidt, G. (1979) J. Cell. Biol. 81, 461-483.
Cline, K., Andrews, J., Mersey, B., Newcomb, E.H. and Keegstra, K. (1981) Proc. Natl. Acad. Sci. USA 78, 3595-3599.
Dorne, A.-J., Block, M.A., Joyard, J. and Douce, R. (1982) FEBS lett. 145, 30-34.
Dorne, A.-J., Joyard, J., Block, M.A. and Douce, R. (1985) J. Cell Biol. 100, 1690-1697.
Douce, R. (1974) Science 183, 852-853.
Douce, R. and Guillot-Salomon, T. (1970) FEBS Lett. 11, 121-126.
Douce, R. and Joyard, J. (1980) in: The Biochemistry of Plants, volume 4, P.K. Stumpf, ed., pp. 321-362. Academic Press, New-York.
Douce, R., Holtz, R.B. and Benson, A.A. (1973) J. Biol. Chem. 248, 7215-7222.
Douce, R., Block, M.A., Dorne, A.-J. and Joyard, J. (1984) Subcellular Biochem. 10, 1-84.
Eichenberger, W. (1982) Plant Sci. Lett. 24, 91-95.
Fishwick, M.J. and Wright, A.J. (1980) Phytochemistry 19, 55-59.
Frentzen, M., Heinz, E., McKeon, T.A. and Stumpf, P.K. (1983) Eur. J. Biochem. 129, 629-636.
Harwood, J.L. and Russell, N.J. (1984) Lipids in Plants and Microbes. George Allen and Unwin, London.
Heber, H. and Heldt, H.W. (1981) Ann. Rev. Plant Physiol. 32, 139-168.
Heemskerk, J. (1986) Ph.D. Thesis. University of Nijmegen, Holland.
Heemskerk, J., Wintermans, J.F.G.M., Joyard, J., Block, M.A., Dorne, A.-J. and Douce R. (1986) Biochim. Biophys. Acta 877, 281-289.
Heinz, E. (1977) in: Lipids and Lipid Polymers in Higher Plants, M. Tevini and H.K. Lichtenthaler, eds., pp. 102-120. Springer-Verlag, Berlin.
Heinz, E. and Roughan, P.G. (1983) Plant Physiol. 72, 273-279.

Jelsema, C.L., Michaels, A.S., Janero, D.R. and Barrnett, R.J. (1982) J. Cell Sci. 58, 469-488.
Journet, E.-P. and Douce, R. (1985) Plant Physiol. 79, 458-467.
Joyard, J. and Douce, R. (1976) C.R. Acad. Sci. Paris 282, 1515-1518.
Joyard, J. and Douce, R. (1977) Biochim. Biophys. Acta 486, 273-285.
Joyard, J. and Douce, R. (1979) FEBS Lett. 102, 147-150.
Joyard, J., Chuzel, M. and Douce, R. (1979) In: Advances in the Biochemistry and Physiology of Plant Lipids, L.-A. Appelqvist and C. Liljenberg, eds., pp. 181-186. Elsevier, Amsterdam.
Joyard, J., Blée, E. and Douce, R. (1986) Biochim. Biophys. Acta in press (BBA 52409).
Kleppinger-Sparace, K.F., Mudd, J.B. and Bishop, D.G. (1985) Arch. Biochem. Biophys. 240, 859-865.
Liedvogel, B., and Kleinig, H. (1070) Planta 144, 467-471.
Lefort-Tran, M. (1981) Ber. Dtsch, Bot. Ges. 94, 463-476.
Lewin, R.A. (1981) in: Origins and evolution of eukaryotic intracellular organelles, J.F. Fredrick, ed., pp. 325-329. Ann. N.Y. Acad. Sci. New-York.
Lucy, J.A. and Ahkong, Q.J. (1986) FEBS Lett. 199, 1-11.
Mendiola-Morgenthaler, L., Eichenberger, W. and Boschetti, A. (1985) Plant Sci. Lett. 41, 97-104.
Michaels, A.S., Jelsema, C.L. and Barrnett, R.J. (1983) J. Ultra. Res. 82, 35-51.
Miquel, M., Block, M.A., Joyard, J., Dorne, A.-J., Dubacq, J.-P., Kader, J.C. and Douce, R. (1984) In: Structure, Function and Metabolism of Plant Lipids P.A. Siegenthaler and W. Eichenberger, eds., pp. 295-298. Elsevier, Amsterdam.
Mudd, J.B. and de Zacks, R. (1981) Arch. Biochem. Biophys. 209, 584-591.
Murata, N. and Sato, N. (1982) in: Biochemistry and Metabolism of Plant Lipids, J.F.G.M. Wintermans and P.J.C. Kuiper, eds., pp. 165-168. Elsevier Biomedical Press, Amsterdam.
Murata, N. and Sato, N. (1983) Plant Cell Physiol. 24, 133-138.
Murata, N., Sato, N., Omata, T. and Kuwabara, T. (1981) Plant Cell Physiol. 22, 855-866.
Nichols, B.W., Harris, R.V. and James, A.T. (1965) Biochem. Biophys. Res. Commun. 20, 256-262.
Omata, T. and Murata, N. (1983) Plant Cell Physiol. 24, 1101-1112.
Peschek, G.A. (1984) Subcellular Biochem. 10, 85-191.
Roughan, P.G. and Slack, C.R. (1982) Annu. Rev. Plant Physiol. 33, 97-132.
Sato, N. and Furuya, M. (1984) in: Structure, Function and Metabolism of Plant Lipids, P.A. Siegenthaler and W. Eichenberger, eds., pp. 171-174, Elsevier, Amsterdam.
Sato, N. and Murata, N. (1982a) Biochim. Biophys. Acta 710, 271-278.
Sato, N. and Murata, N. (1982b) Biochim. Biophys. Acta 710, 279-289.
Sato, N. and Murata, N. (1982c) Plant Cell. Physiol. 23, 1115-1120.
Siebertz, H.P., Heinz, E., Linscheid, M., Joyard, J. and Douce, R. (1979) Eur. J. Biochem. 101, 429-438.
Sparace, S.A. and Mudd, J.B. (1982) Plant Physiol. 70, 1260-1264.
Stumpf, P.K. (1980) in: The Biochemistry of Plants, Vol. 4, P.K. Stumpf, ed., pp. 177-204, Academic Press, New-York.
Van Besouw, A. and Wintermans, J.F.G.M. (1978). Biochim. Biophys. Acta 529, 44-53.

REGULATION OF CHLOROPLAST DIFFERENTIATION: COOPERATION BETWEEN LIGHT-INDUCED PROCESSES AND INTERNAL ADAPTATION

Peter Brandt, Jörg Winter, Barbara von Kessel and Beatrix Kohnke

Abteilung für Experimentelle Phykologie, Pflanzenphysiologisches Institut der Universität Göttingen, Untere Karspüle 2, D-3400 Göttingen, F.R.G.

(Supported by grant Br 652/5-5 from the DFG)

The development of plants depends on the harmonic cooperation of three genetic compartments: the nucleocytoplasm, plastids and mitochondria. One single DNA molecule contains the genetic information of the plastid. This chromosome is known to encode part of the plastidial transcription and translation machineries, and some of the components involved in the photosynthetic process. The photosynthetic membrane is composed of about 50 polypeptide species, most of which are found in seven multisubunit complexes (von Wettstein 1980). These are the photosystem I (PSI) - composed of the reaction center (RCI) and the PSI antenna system (ASI) -, the light-harvesting system LHCI (related to PSI), the photosystem II (PSII) - composed of the reaction center (RCII) and the PSII antenna system (ASII) -, the light-harvesting system LHCII (generally related to PSII), the water-splitting-complex (OEC), the cytochrome b/f-complex (Cyt b/f) and the ATPase - composed of the intrinsic part CF_o and the extrinsic part CF_1. Approximately half of these proteins are encoded by the plastid chromosome, including all of the chlorophyll a-containing reaction center apoproteins and cytochrome, and half in the nucleus (Herrmann et al. 1985).

During the last three years we have isolated four of the seven thylakoid complexes from the unicellular alga *Euglena gracilis* and we have characterized their properties. Our main interest, however, has been the elucidation of chloroplast differentiation in *Euglena*. In our case this means the development of 'young' chloroplasts, after cell division, into 'old' chloroplasts just before cell division. The experimental approach for these investigations were mass cultures of *Euglena gracilis*, which were synchronized by a light-dark change of 14 hours light and 10 hours dark. It is important to mention that the synchronisation of *Euglena* mass cultures was always done in an autotrophic culture medium together with vitamins B_1 and B_{12}, because chloroplast differentiation in *Euglena gracilis* is affected by organic C-sources (Brandt 1981).

There are 11 chloroplasts in autotrophically cultured *Euglena* cells which are located in the periphery of the cell. Chloroplast biogenesis in synchronized *Euglena gracilis* cultures takes place during the light-time of the cell cycle, but neither the chloroplasts nor the *Euglena* cells divide during this period (Brandt and von Kessel 1983). It is important that a developmental period for the chloroplasts is provided by the light-dark change in which division of the cells or of the chloroplasts does not occur (Fig.1). Total protein and chlorophyll of *Euglena gracilis* double in a linear fashion during this light-time (Buetow et al. 1980). Synthesis of thylakoid components, however, takes place only at certain stages during the cell cycle and in a characteristic step-by-step fashion (Fig.2). Synthesis of CPI, which is related to PSI, occurs during the 1st to the 2nd hour of the light-time and is obviously light-dependent. Synthesis of cytochrome f, however, begins just after the division of the chloroplasts during the last part of the dark-time and ends at the beginning of the light-time. Cytochrome b-563, which is the other main constituent of the cytochrome b/f-complex, is synthesized later on

Algal Development (Molecular and Cellular Aspects)
Ed. W. Wiessner, D.G. Robinson, and R.C. Starr

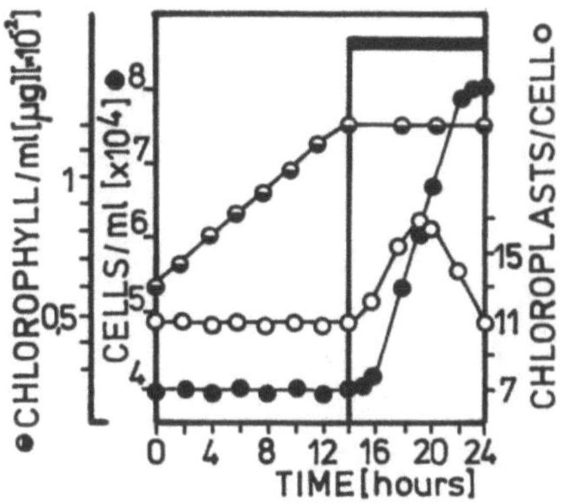

Fig.1. Cells/ml(●),chlorophyll/ml(⊖) and chloroplasts/cell(○) of a Euglena gracilis culture synchronized by a light-dark change of 14 h light and 10 h dark.(Brandt and von Kessel 1983)

in the second part of the light-time. At this stage of the cell cycle the synthesis of CPa, which is related to PSII, and the assembly of LHCII also occur. This LHCII is the only light-harvesting system in Euglena chloroplasts,the LHCI related to PSI is missing (Brandt 1980; Cunningham and Schiff 1986).

This step-by-step differentiation of the thylakoid system in Euglena chloroplasts causes drastic changes in photosynthetic efficiency (Brandt 1981; Brandt and von Kessel 1983) and minor changes in the ultrastructure in the chloroplasts (Winter and Brandt 1986). In particular photosynthetic oxygen evolution as well as PSI- and PSII-activities are low at the 6th and at about the 12th hour of the light-time. In the chloroplasts of Euglena gracilis distinct grana-stacks are missing. In limitted regions of the thylakoid system at the most only two or three thylakoids are found stacked. In the 'end-regions' of the chloroplasts, however, regions of high stacking, called grana-homologs, are seen (Fig.3)(Winter and Brandt 1986). They disappear in the early part of the light-time, when newly-synthesized CPI is inserted into the thylakoid system of the Euglena chloroplasts (Fig.2). The basic function of the LHCII is to absorb light and to transfer excitation energy to the two photosystems (Butler 1978). The required functional association of LHCII to PSII or PSI is regulated by phosphorylation or dephosphorylation (Allen et al. 1981), which has been demonstrated in higher plant chloroplasts (Bennett 1977; Steinbeck et al. 1979) in chloroplasts of Chlamydomonas (Owens and Ohad 1982, 1983) and in Euglena gracilis (Beliveau and Bellemare 1979). In higher plants

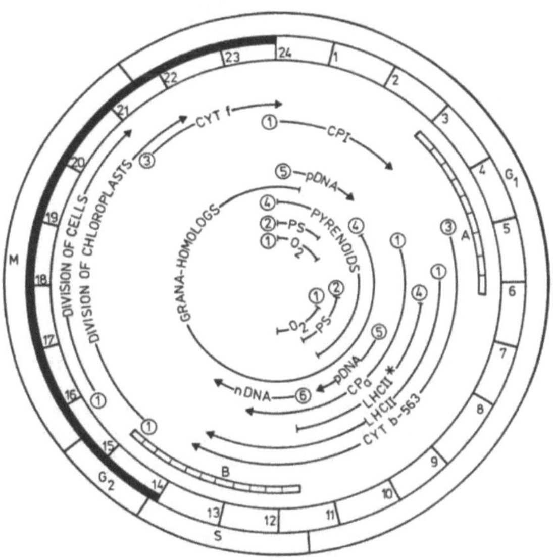

Fig.2. General scheme of step-by-step differentiation of the chloroplasts in Euglena gracilis during the cell cycle. Explanations in the text. (1)Brandt and von Kessel 1983;(2)Brandt 1981;(3)Kohnke and Brandt 1984;(4)Winter and Brandt 1986; (5)Brandt 1975;(6)Edmunds 1964;A and B = temperature sensitive stages (Brandt and Wiessner 1977)

this phosphorylation causes a decrease of thylakoid stacking (Staehelin and Arntzen 1983). In the unicellular alga Euglena gracilis, however, such structural alterations could not be observed, although phosphorylation of thylakoid proteins (Beliveau and Bellemare 1979)

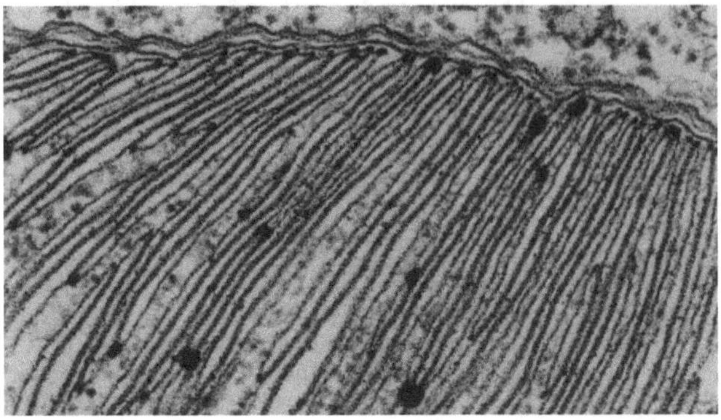

Fig.3. Electron micrograph of a section through the region of grana-homologs in a chloroplast of Euglena gracilis at the 8th hour of the dark-time of the cell cycle. (Winter and Brandt 1986)(x110.000)

is similar to that of higher plant chloroplasts (Bennett 1977). Therefore we must investigate whether the functional association of LHCII with PSI or with PSII occurs in Euglena chloroplasts without any distinct intrachloroplastic structural changes. Furthermore the functional integration of newly-synthesized CPI and of the newly-synthesized LHCII into the existing photosynthetic apparatus has to be examined. The experimental proof of the functional association of LHCII with PSII has been demonstrated by fluorescence measurements in the presence of DCMU after preillumination with light of 725 nm ('light I') or with light of 620 nm ('light II') or with white light (Fig.4). Samples were taken from the synchronized Euglena culture every hour during the light-time. The variable fluorescence F_v yield of light I samples was always above, and that of the light II samples was always below the F_v of the control. The ratio of F_v from the light I samples (A) to the F_v from the light II samples (B) changes during the light-time of the cell cycle of Euglena gracilis in a characteristic manner (Fig.5). It increases after the 5th hour of the light-time and has a maximum at the 7th hour. These data imply that under appropriate light conditions a part of LHCII available at about the 7th hour is not permanently related to PSII at this stage in the Euglena gracilis life cycle. This time corresponds to the insertion of newly assembled LHCII into the thylakoids (Fig.2).
Evidence for a functional association of newly assembled LHCII with PSI is demonstrated by the simultaneous measurement of the fluorescence kinetics of PSI and of PSII at 77 K (Fig.6). From these measurements the ratio of the fluorescence kinetics of PSI and of PSII under light I or light II conditions was calculated. The cell cycle-dependent change of these ratios can easily be demonstrated by a plot of the calculated values of the ratio of the fluorescence kinetics of PSII under light I or light II conditions (abscissa) in dependence of the corresponding values of PSI (ordinate). The result is always an ascending straight Butler-line (1978)(Fig.6). The increased angle x of this straight line represents graphically the degree of energy distribution between PSI and PSII under light I or light II conditions: the steeper the slope the better is the energy supply of PSI. The cell cycle-specific change of this energy distribution during cell development, caused by the preillumination with light I or with light II, is presented by comparison of the increase of the two corresponding straight lines (Fig.7). The angular difference $X_{II}-X_{I}$ of two consecutive measurements is taken as an index for the cell cycle-specific change of the energy distribution in the photosynthetic apparatus of Euglena gracilis, i.e. the grade of the functional association of the LHCII with PSII or with PSI. The resulting graph confirmed the assertion of the measurements of the fluorescence induction curves (Fig.5).

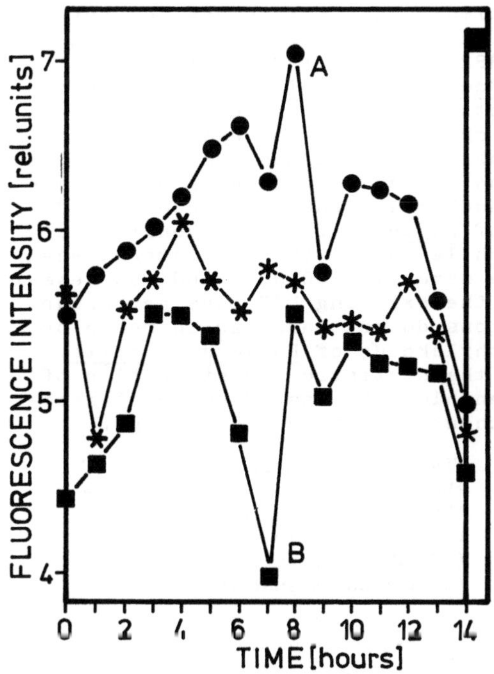

Fig.4. Changes of the variable fluorescence F_v at room temperature of Euglena cells during the light-time of the cell cycle after illumination of the cells for 15 min with white light (✱) or with light of 725 nm (●) or with light of 620 nm (■).(Winter and Brandt 1986)

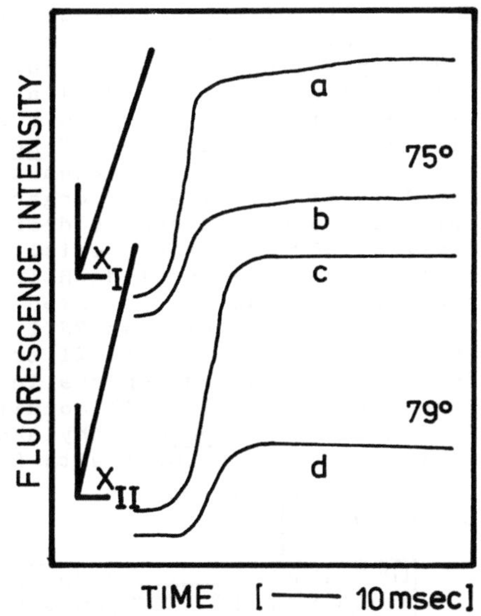

Fig.6. Changes of the fluorescence kinetics at 77 K of PSI after preillumination with 725 nm-light(a) or with 620 nm-light(c) and of PSII after preillumination with 725 nm-light (b) or with 620 nm-light(d) of Euglena cells at the 8th hour of the cell cycle.Further details in the text.(Winter and Brandt 1986)

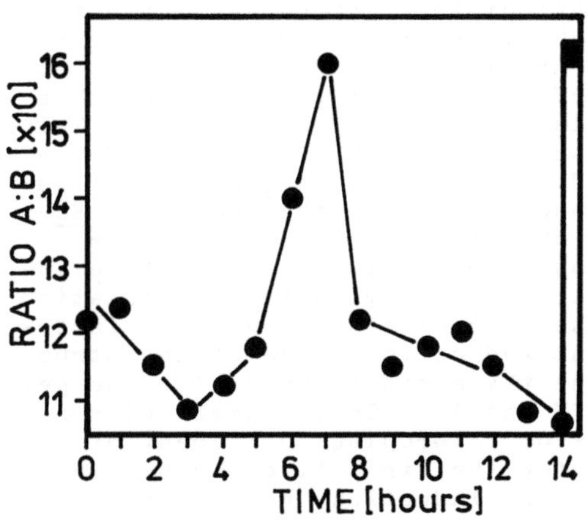

Fig.5. Changes of the ratio variable fluorescence F_v after 725 nm-illumination(A)/ variable fluorescence F_v after 620 nm-illumination(B) of Euglena cells during the light-time of the cell cycle. (Winter and Brandt 1986)

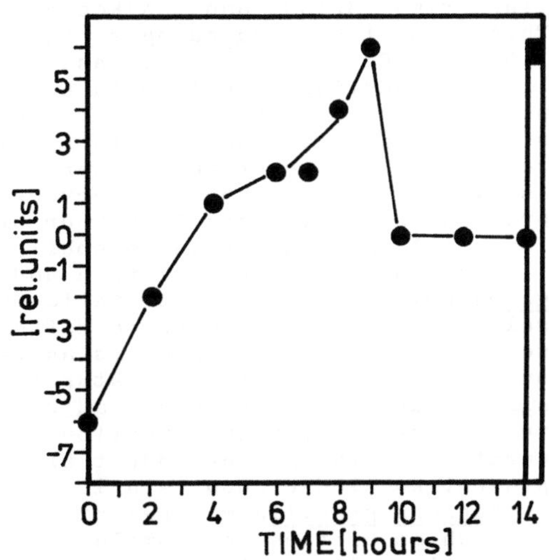

Fig.7. Changes of the angle differences (see Fig.6), which were calculated from the fluorescence kinetics of PSI and of PSII of Euglena cells during the light-time of the cell cycle (X_{II}-X_I). (Winter and Brandt 1986)

The results of these investigations imply that the theory for the structural and functional organization of the thylakoid system in higher plant chloroplasts (Barber 1982; Biggins 1982; Kyle et al. 1983; Steinback et al 1979) and in Scenedesmus obliquus (Krupinska et al. 1985) is valid in certain respects for the thylakoid differentiation during the ontogeny of Euglena chloroplasts. Nevertheless, there are also differences between these objects. In opposition to the general model of thylakoid organization, which postulates a direct correlation between stacking decrease and mobile LHCII (Staehelin and Arntzen 1983), in the special case of Euglena it can be considered that (a) newly assembled LHCII is mobile and/or (b) the lateral segregation of the CPa inserted at the same time (Brandt and von Kessel 1983) and functioning as PSII, from the existing PSI does not occur. Therefore, state I-state II transitions do not cause an increase or decrease of stacked regions in the thylakoids of Euglena gracilis. Moreover, under light II conditions the statistical distribution of PSII and PSI in Euglena thylakoids enables the fast functional association of mobile LHCII with PSI without any structural change.

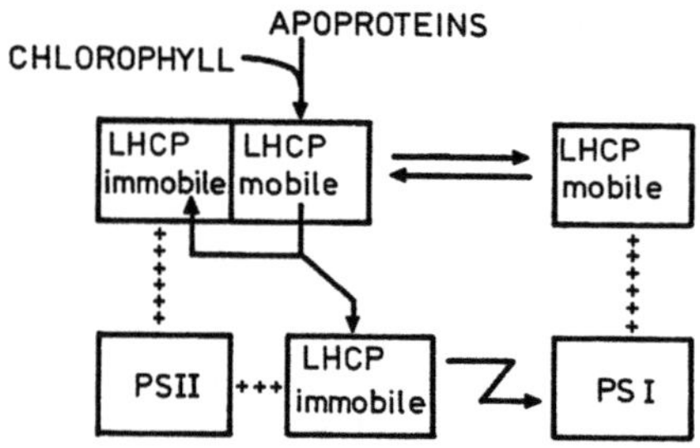

Fig.8. The 'course of career' of the newly assembled LHCII in the thylakoids of Euglena gracilis. (+), functionally organized with; (→), energy transfer. (Winter and Brandt 1986)

The possible 'course of career' of the LHCII in Euglena chloroplasts is summarized in Fig.8. LHCII assembled during the second part of the light-time (Brandt and von Kessel 1983)(Fig.2) is of mobile nature. This means that, according to the hypothesis of Allen and Holmes (1986), LHCII can be either associated as an additional peripheral light-harvesting complex with an intermediate light-harvesting complex, which is tightly bound to the RCII, or drifts away from the PSI after phosphorylation of both LHCIIs and is then functionally connected to PSI. The continued insertion of CPa, however, means that this LHC is now immobilized by irreversible binding to the new RCIIs. This process ends after the 10th hour of the light-time. After this developmental stage there is no mobile LHCII present in Euglena gracilis.

Obviously the organization and function of the thylakoids in Euglena chloroplasts is not static but dynamic in nature during the cell cycle. The step-by-step differentiation includes consecutive insertion of various components as well as a concomitant new adaptation of the photosynthetic apparatus to the altered situation. This interplay ensures the continuous adjustment and optimization of the structure, organization and function of the thylakoid membrane for the generation of the proper amount of NADPH and ATP. On the basis of our present knowledge about single events of chloroplast differentiation in Euglena gracilis we present here, for the first time, an hypothesis, which unifies the complex network of induction, regulation and adaptation. It is without saying that in doing so more questions will be raised than can be answered.

(1)It is now evident for all plastome coded protein species that their mRNAs are present in the light and in the dark (Herrmann et al. 1985). In the special case of Euglena gracilis this has been demonstrated for the chlorophyll-binding proteins by Devic and Schantz (1984). We have been able to demonstrate the same phenomenon for the

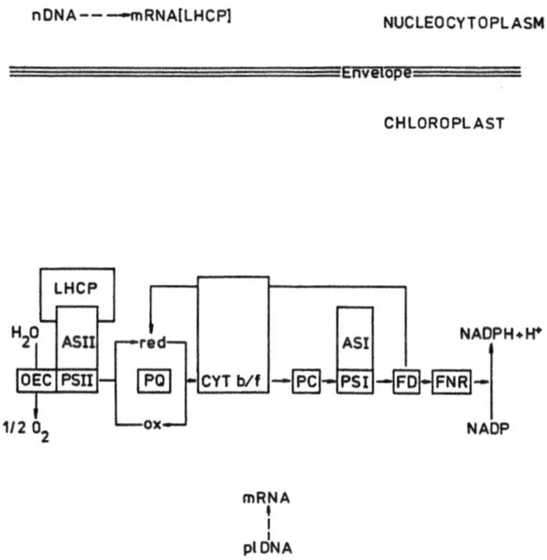

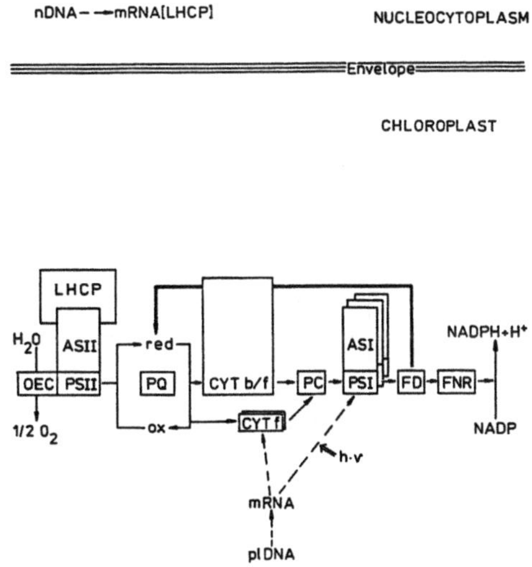

Fig.9a. Schematic presentation of the situation of the photosynthetic apparatus of Euglena chloroplasts near after the chloroplast division (7th hour of the dark-time of the cell cycle)

Fig.9b. Schematic presentation of the situation of the photosynthetic apparatus of Euglena chloroplasts in the stage 7th h of the dark-time = 3rd h of the light-time of the cell cycle

mRNA of the LHCII-apoprotein which is also encoded in the nucleus (Brandt and Kloppstech, unpublished). In general the occurrence of newly-synthesized proteins in the thylakoid system of Euglena chloroplasts is under post-transcriptional or translational control.

(2)After chloroplast division and cell division the 'young' chloroplasts contain a complete photosynthetic apparatus (Fig.9a).

(3)The first events in the last part of the dark-time and in the first part of the light-time are the synthesis of cytochrome f and the subsequent synthesis of CPI, i.e. of PSI and ASI (Fig.9b). CPI-synthesis is obviously light-induced, but the induction mechanism for cyt f-synthesis is unknown. We have previously demonstrated that this cytochrome f is much more extrinsic than that bound in the cytochrome b/f-complex (Kohnke and Brandt 1984). This view of cytochrome f location in the thylakoids of Euglena corresponds with immunological studies, which show that that part of cytochrome f, which is involved in linear electron transport, is located on the outer surface of the mem brane, whereas the cytochrome involved in the cyclic phosphorylation seems to reside within the membrane (Schmid et al. 1977). Both the increased amount of PSI and the elevated amount of cytochrome f may cause an increased cyclic phosphorylation. The ratio of ATP/NADPH is also elevated and much more of the plastochinone is in the reduced stage (Fig.9b).

(4)After the 6th hour of the light-time certain proteins are incorporated into the envelope (Brandt 1982) which enable the translocation of the now translated LHCII-apoprotein (Fig.9c). The higher ATP level and the amount of reduced plastochinone favour the functional association of the new LHCII with PSI probably by phosphorylation (Allen et al. 1981).

(5)At the same time in the cell cycle, cytochrome b-563 and CPa are synthesized and inserted into the photosynthetic apparatus (Fig.9d). Therefore the available amount of PSII increases as does the plasto-

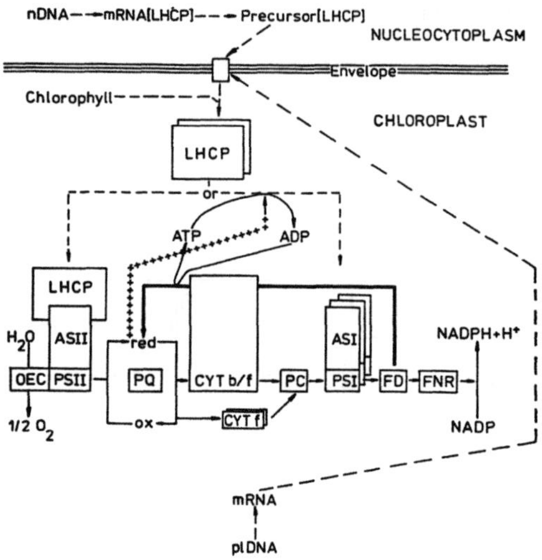

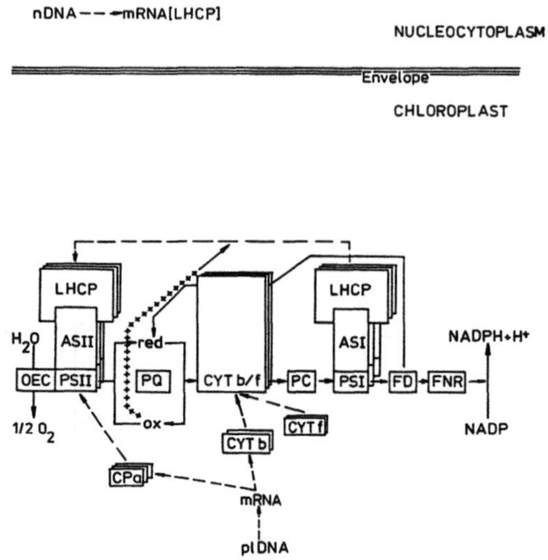

Fig.9c. Fig.9d.
Schematic presentation of the situation of the photosynthetic apparatus of *Euglena* chloroplasts in the stage between the 6th and the 9th hour of the light-time of the cell cycle. (+), activation of catalysing enzymes.

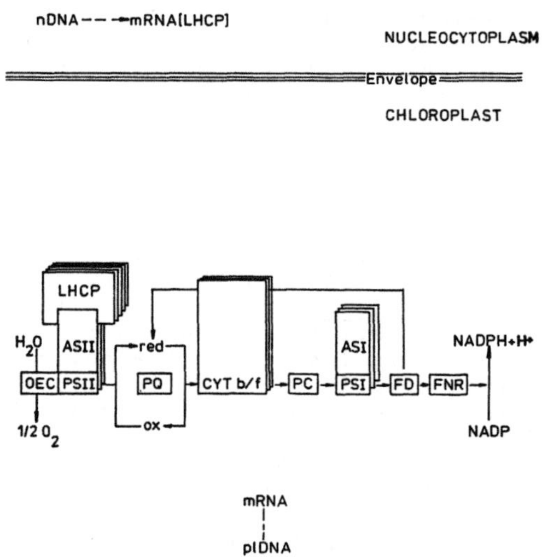

Fig.9e. Schematic presentation of the situation of the photosynthetic apparatus of *Euglena* chloroplasts at the end of chloroplast differentiation at about the 14th hour of the light-time of the cell cycle

chinone pool in the oxidized stage. According to the common model (Allen et al. 1981) the mobile LHCII associated with PSI will be re-organized with respect to PSII. The newly-synthesized cytochrome b-563 will be assembled with the already present cytochrome f to cy-tochrome b/f-complexes.

(6)In the 'old' chloroplasts at the end of the light-time (Fig.9e) the photosynthetic apparatus of *Euglena* chloroplasts will be enlarged but the stoichiometries of the thylakoid components will be identical with the situation at the beginning of chloroplast differentiation (Fig.9a).

It is obvious that this proposal of chloroplast differentiation will be modified in the future, when more details about structure-function relationship as well as about the insertion time of the outstanding components will be available. The present proposal, however, clearly shows that chloroplast differentiation is mainly regulated at the post-transcriptional level together with other mechanisms such as the ratio of ATP/NADPH, the presence of certain thylakoid components, which are prerequisite for the insertion of the subsequent proteins, and internal adaptation of the photosynthetic apparatus, i.e. opti-mizing the system by re-arrangements.

Acknowledgement

We thank Prof. D.G.Robinson for his gracious efforts on our behalf.

References

Allen,J.F.,Holmes,N.G.(1986)FEBS Letters 202,175-181
Allen,J.F.,Bennett,J.,Steinback,K.E.,Arntzen,C.J.(1981)Nature 291,25
Beliveau,R.,Bellemare,G.(1979)Biochem.Biophys.Res.Commun.88,797-8o3
Bennett,J.(1977)Nature 269,344-346
Brandt,P.(1975)Planta (Berl.) 124,105-107
Brandt,P.(1980)Z.Pflanzenphysiol. 100,95-105
Brandt,P.(1981)Ber.Deutsch.Bot.Gesell. 94,419-434
Brandt,P.(1981)Z.Pflanzenphysiol. 103,65-74
Brandt,P.(1982)Plant Science Lett. 28,237-244
Brandt,P.,von Kessel,B.(1983)Plant Physiol. 72,616-619
Brandt,P.,Wiessner,W.(1977)Z.Pflanzenphysiol. 85,53-60
Brandt,P.,Winter,J.(1987)Protoplasma (in press)
Buetow,D.E.,Wurtz,E.A.,Gallagher,T.(1980)In: Nuclear-Cytoplasmic Interactions in the Cell Cycle.ed.Whitson,G.L.,Academic Press, New York,pp.9-55
Butler,W.L.(1978)Annu.Rev.Plant Physiol. 29,345-378
Cunningham,F.X.,Schiff,J.A.(1986)Plant Physiol. 80,223-230
Devic,M.,Schantz,R.(1984)In: Advances in Photosynthesis Research.ed. Sybesma,C.,Nijhoff/Junk,The Hague,Vol.IV,pp.575-578
Edmunds,L.N.Jr.(1964)Science 145,266-268
Herrmann,R.G.,Westhoff,P.,Alt,J.,Tittgen,J.,Nelson,N.(1985)In: Mole-cular Form and Function of the Plant Genome.eds.v.Vloten-Dolting,L., Groot,G.,Hall,T.,Plenum Publ.Corp.,pp.233-256
Kohnke,B.,Brandt,P.(1984)Biochim.Biophys.Acta 766,156-160
Owens,G.C.,Ohad,I.(1982)J.Cell Biol. 93,712-718
Owens,G.C.,Ohad,I.(1983)Biochim.Biophys.Acta 722,234-241
Schmid,G.H.,Radunz,A.,Menke,W.(1977)Z.Naturforsch. 32c,271-280
Staehelin,L.A.,Arntzen,C.J.(1983)J.Cell Biol. 97,1327-1337
Steinback,K.E.,Mullet,J.E.,Arntzen,C.J.(1979)Plant Physiol. 63,S-27
von Wettstein,D.(1980)In: International Cell Biology.ed.Schweiger,H. G.,Springer-Verlag,Berlin,pp.250-272
Winter,J.,Brandt,P.(1986)Plant Physiol. 81,548-552

INTERACTION OF NUCLEAR AND CHLOROPLAST MUTATIONS IN BIOGENESIS OF CHLOROPLAST RIBOSOMES IN *CHLAMYDOMONAS*

E.H. Harris[1], J.E. Boynton[1], and N.W. Gillham[2]
Departments of Botany[1] and Zoology[2]
Duke University
Durham, NC 27706

In *Chlamydomonas*, as in higher plants, the chloroplast genome codes for chloroplast ribosomal RNAs, tRNAs, about 1/3 of the ribosomal proteins, and a number of proteins of the photosynthetic apparatus (Table 1, Fig. 1). Other proteins are coded by the nuclear genome and imported, generally as precursors which are processed to mature form within the chloroplast (Table 1). Because *C. reinhardtii* can dispense with photosynthetic functions if provided acetate as an alternative carbon source, this species is an ideal model system for investigation of mutations affecting these components. In the case of the large subunit of ribulose bisphosphate carboxylase (RUBISCO; Spreitzer & Ogren, 1985), the 32 kd herbicide binding protein (Erickson et al., 1985), and the beta subunit of the CF_0CF_1 ATPase (Woessner et al., 1986), analysis of chloroplast genes has extended to the DNA sequence level, and to comparison of mutations and reversions within the same gene.

One major aspect of research in our laboratory is the investigation of chloroplast ribosome biogenesis (for reviews see Harris et al., 1976; Davidson et al., 1978; Boynton et al., 1980; Schmidt et al., 1985). This paper reviews the current status of this work and includes new data on the molecular alterations in antibiotic resistance mutations.

Mutations which block assembly of chloroplast ribosomes (Table 2) prevent efficient synthesis of components listed at the top of Table 1, and thereby result in a syndrome of defects, including loss of photosynthetic electron transport, photophosphorylation, and RUBISCO activity, and striking alterations in chloroplast lamellar structure (Goodenough & Levine, 1970; Boynton et al., 1972; Harris et al., 1974). Some of these mutants fail to assemble both ribosomal subunits, while others accumulate 54s particles which appear to be intact large subunits. The specific ribosomal component affected is known only for the nuclear *cr-6* locus. Myers et al. (1984) showed that the allelic mutations *cr-6* and *cr-7*

Algal Development (Molecular and Cellular Aspects)
Ed. W. Wiessner, D.G. Robinson, and R.C. Starr

TABLE 1. Components of the chloroplast whose sites of coding and/or synthesis have been identified in *C. reinhardtii*.

Chloroplast components coded by chloroplast genes, and synthesized on chloroplast ribosomes:

tRNAs: at least 12	Bergmann et al. (1985)
rRNAs: 23s, 16s, 7s, 5s, 3s	Rochaix & Malnoe, 1978
ribosomal proteins: 14 proteins of small subunit 5 or 6 proteins of large subunit	Schmidt et al.(1983,1985)
EF_{Tu}	Watson & Surzycki (1982)
CF_0CF_1 ATPase: alpha, beta, epsilon, I, III, IV subunits	Woessner et al. (1986), (1987)
RUBISCO large subunit	Malnoe et al. (1979); Dron et al. (1982)
thylakoid membrane proteins: at least 9, including: 51 kd chl a apoprotein P_{700} chl a apoproteins 1a, 1b 44 kd chl a apoprotein 32 kd herbicide binding protein D1 32 kd protein D2	Chua & Gillham (1977), Delepelaire (1984), Wollman et al. (1980), Rochaix et al. (1984)
cytochrome b_6	Garnier & Maroc (1984)
cytochrome f	Delepelaire (1983)

Chloroplast components coded by nuclear genes, and synthesized on cytoplasmic ribosomes:

ribosomal proteins: 17 proteins of small subunit 27 or 26 proteins of large subunit	Schmidt et al. (1983)
CF_0CF_1 ATPase: gamma, delta, II	Woessner et al. (1986)
RUBISCO small subunit	Goldschmidt-Clermont (1986) Marks et al. (1986), Mishkind et al. (1985)
phosphoribulokinase	Salvucci & Ogren (1985)
thylakoid membrane proteins: at least 24, including: proteins of light harvestng and water splitting complexes chl a/b binding protein 34 kd protein	Chua & Gillham (1977), Delepelaire (1984), Wollman et al. (1980)
ferredoxin	Huisman et al. (1979)
plastocyanin	Merchant & Bogorad (1986)
cytochrome c-552	Merchant & Bogorad (1986)
(most other soluble enzymes are also assumed to be coded in the nucleus)	

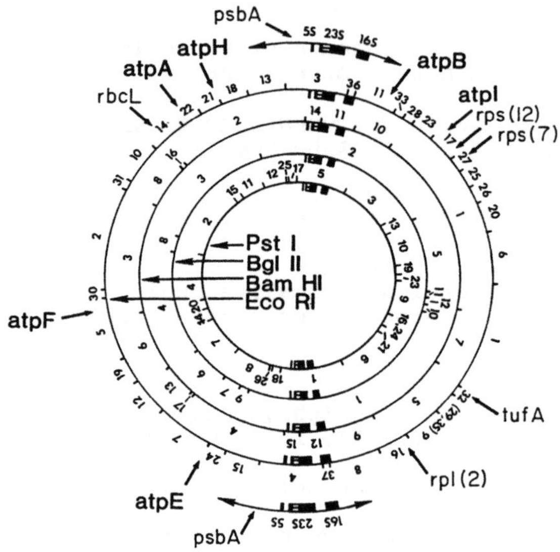

FIGURE 1: Map of chloroplast genome of *C. reinhardtii*. Restriction map by Rochaix (1978), redrawn by Grant et al. (1980); gene positions from Malnoe et al. (1979; Watson & Surzycki (1982); Waddell et al. (1984); Erickson et al. (1984); Schmidt et al. (1985); Vallet & Rochaix (1985); Woessner et al. (1986, 1987).

resulted in loss of two proteins of the large ribosomal subunit, L-13, made in the chloroplast, and L-29, made in the cytoplasm. L-13 was synthesized in the *cr-6* and *cr-7* mutants, but not assembled, presumably as a secondary consequence of loss of L-29.

TABLE 2. Mutations blocking chloroplast ribosome assembly and function. From Boynton et al. (1972); Harris et al. (1974); Gillham et al. (1976); Shepherd et al. (1979); Myers et al. (1984).

Mutant locus	Genome	Ribosomal phenotype
ac-20	nuclear	loss of both subunits
ac-u-d	chloroplast	accumulation of 54s particles
ac-u-h	chloroplast	accumulation of 54s particles
cr-1	nuclear	accumulation of 54s particles
cr-2	nuclear	accumulation of 54s particles
cr-3	nuclear	accumulation of 54s particles
cr-4	nuclear	loss of both subunits
cr-5	nuclear	accumulation of 54s particles
cr-6/cr-7	nuclear	large subunit incompletely formed improper assembly of monomers

Mutants of *C. reinhardtii* resistant to several antibiotics which block protein synthesis in bacteria have been isolated (Table 3). Most of these show the uniparental pattern of inheritance characteristic of chloroplast genes (cf. Gillham, 1978; Grant et al., 1980), and have been mapped by recombination analysis in a single linear linkage group (Harris et al., 1977; Bartlett et al., 1979). A nuclear streptomycin resistance locus and two erythromycin resistance loci affecting the chloroplast ribosomes are also known (Harris et al., 1977; Davidson et al., 1978). In each case the ribosomal subunits affected by these resistance mutations are the same as those affected by similar mutations in bacteria (Table 3). Only the ery-1 locus has been shown definitively to be a structural gene for a ribosomal protein (L-6; Davidson et al., 1974).

TABLE 3. Antibiotic resistance mutations in C. reinhardtii (cf. Sager, 1972; Schlanger & Sager, 1974; Bartlett et al., 1979; Davidson et al., 1978).

Mutant locus	Genome	Antibiotic resistance	Subunit affected
sr-1	nuclear	streptomycin	small
sr-u-2-60	chloroplast	streptomycin	small
sr-u-2-23	chloroplast	streptomycin	small
sr-u-sm3	chloroplast	streptomycin	small
sr-u-sm2	chloroplast	streptomycin	small
nr-u-2-1	chloroplast	neamine/kanamycin	small
spr-u-1-6-2	chloroplast	spectinomycin	small
ery-1	nuclear	erythromcyin	large
ery-2	nuclear	erythromycin	large
er-u-1a	chloroplast	erythromycin	large
er-u-37	chloroplast	erythromycin	large

Recent work in our laboratory has centered on identifying genes for chloroplast ribosomal proteins, and on determining the molecular basis for chloroplast antibiotic resistance. By heterologous hybridization with probes from spinach, maize, and Euglena, regions of the chloroplast genome which appear to code for proteins homologous to E. coli L-2, S-7 and S-12 were identified (Fig. 1).

Prompted by the findings of Sigmund et al. (1984) and Fromm et al. (1985) that spectinomycin resistance mutations in E. coli and tobacco, respectively, alter a restriction site for the enzyme Aat II in the gene for the 16s rRNA, we screened chloroplast DNA from our collection of spectinomycin resistant mutants of Chlamydomonas with this enzyme. All 13 of the non-conditional mutants examined have lost this cut site in the 16s rRNA gene (Fig. 2, Table 4), whereas this site is unaltered in three resistant mutants that are also temperature-sensitive conditional lethals. The latter, which have not yet been mapped genetically, may prove to be mutations in a chloroplast gene affecting a ribosomal protein, possibly analogous to the S-5 protein of E. coli (Bollen et al., 1969). The mutants nr-u-2-1 and kr-u-24-2, which are resistant to neamine and kanamycin but sensitive to spectinomycin, are extremely closely linked genetically to the spectinomycin resistance locus defined by spr-u-1-27-3 and spr-u-1-6-2 (Harris et al., 1977). Both mutants show the wild type Aat II cut site in the 16s rRNA gene (Table 4).

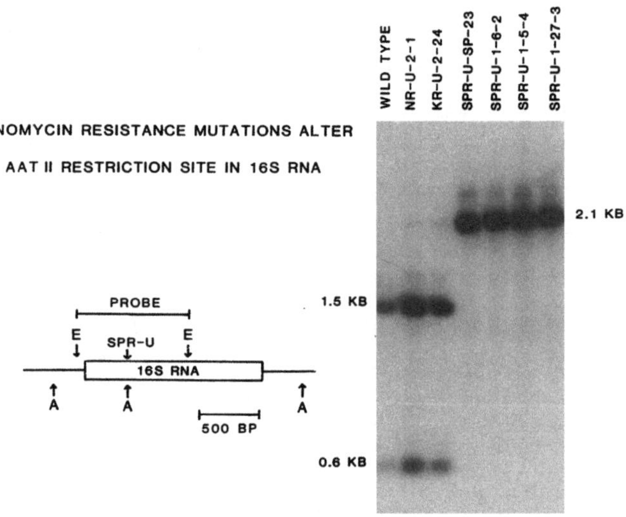

FIGURE 2. Chloroplast DNA from wild type and antibiotic resistant mutants of *C. reinhardtii*, digested with Aat II and probed with a plasmid containing the Eco 36/37 fragment spanning the 3' end of the 16s rRNA gene.

TABLE 4. Effect of certain chloroplast mutations conferring resistance to spectinomycin, neamine or kanamycin on the conserved Aat II cut site (GACGTC) in the chloroplast gene encoding 16s rRNA.

Strain	Resistance	Additional Phenotype	Aat II site
wild type			present
spr-u-1-6-2	spectinomycin		absent
spr-u-sp-23	spectinomycin		absent
spr-u-1-27-3	spectinomycin	(low level)	absent
spr-u-1-H-4	spectinomycin		absent
spr-u-1-5-2-4	spectinomycin		absent
spr-u-1-6	spectinomycin		absent
spr-u-1-13	spectinomycin		absent
spr-u-1-15	spectinomycin		absent
spr-u-1-19	spectinomycin		absent
spr-u-1-24	spectinomycin		absent
spr-u-1-30	spectinomycin		absent
spr-u-1-10-41	spectinomycin		absent
spr-u-1-10-51	spectinomycin		absent
spr-u-4-43	spectinomycin	cold sensitive	present
spr-u-4-15	spectinomycin	heat/cold sensitive	present
spr-u-3-3	spectinomycin	cold sensitive	present
nr-u-2-1	neamine/kanamycin		present
kr-u-2-24	neamine/kanamycin		present

We cloned the chloroplast DNA fragment Eco 36/37 from wild type and three of the spectinomycin resistant mutants in pUC8, transferred these inserts to the phage m13mp18, and used the dideoxy method (Sanger et al., 1977) and synthetic oligonucleotide primers to sequence ca. 200 bp in the region surrounding the Aat II site shown to be altered in the mutants. All three mutants have base pair changes within the six base-pair Aat II recognition sequence, but these differ from one another and from the base pair change demonstrated in *E. coli* by Sigmund et al. (1984) (Table 5). The *spr-u-1-6-2* and *spr-u-sp-23* mutants, showing high level resistance, alter the same base pair with an A->G transition mutation and an A->C transversion, respectively. The *spr-u-1-27-3* mutant, which is resistant to a lower level of spectinomycin (Bartlett et al., 1979), results from a G->A transition in a different base pair.

TABLE 5. Alterations in the sequence of a conserved region of the gene encoding 16s ribosomal RNA from spectinomycin resistant mutants of *E. coli* and *C. reinhardtii*. Bases in lower case are not conserved. Sequence for wild type *E. coli* from Brosius et al. (1978), for spectinomycin resistant *E. coli* from Sigmund et al. (1984), and for wild type *C. reinhardtii* from Dron et al. (1982b).

E. coli wild type	G g G G A T G A C G T C A A G T C A t C A T G
resistant mutant	G g G G A T G A T G T C A A G T C A t C A T G
C. reinhardtii	
wild type	G a G G A T G A C G T C A A G T C A g C A T G
spr-u-1-6-2	G a G G A T G G C G T C A A G T C A g C A T G
spr-u-sp-23	G a G G A T G C C G T C A A G T C A g C A T G
spr-u-1-27-3	G a G G A T G A C A T C A A G T C A g C A T G
	[-----------] Aat II site (G A C G T C)

Mutations at another site in 16s rRNA, about 270 bp away from the spectinomycin site, can lead to streptomycin resistance in *Euglena* (Montandon et al., 1985) and tobacco (Hildebrand & Bourque, 1985). Mutations in the gene encoding the rRNA of the large subunit of yeast mitochondria and of *E. coli* can produce erythromycin resistance (Sor & Fukuhara, 1982,1984; Sigmund et al., 1984). Provisionally we have aligned our genetic map of chloroplast antibiotic resistance mutations in *C. reinhardtii* (Harris et al., 1977) with the physical map of the chloroplast genome, using the spectinomycin resistance site as a point of fixation. If the *sr-u-2-60* locus is assumed to be equivalent to the streptomycin resistance sites of *Euglena* and tobacco, we can speculate that two of the remaining streptomycin resistance loci also fall in the 16s gene, while the two erythromycin resistance loci will be found in the 23s gene (Fig. 3). Selected regions of the 16s and 23s rRNA genes from the appropriate mutants are now being sequenced.

The remaining streptomycin resistance locus, *sr-u-sm2*, is very likely in a gene for a ribosomal protein (Ohta et al., 1975), most probably that encoding the protein homologous to S-12, which can be mutated to streptomycin resistance in *E. coli*. This gene is highly conserved in higher plants and *E. coli*, and has been located in the

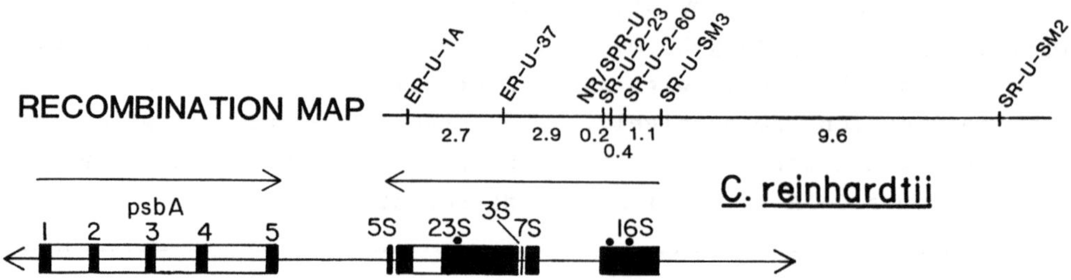

FIGURE 3. Speculative alignment of the recombination map of chloroplast antibiotic resistance genes (Harris et al., 1977) with the physical map of the ribosomal RNA region of the chloroplast genome (Rochaix & Malnoe, 1978). Base pair changes at the _spr-u_ site are documented by sequence analysis (see text), and the position of _sr-u-2-60_ is plotted on the unproven assumption that this locus is equivalent to streptomycin resistance mutations mapped in _Euglena_ and tobacco. Remaining loci are positioned on the basis of their recombination distance from _spr-u_ and _sr-u-2-60_.

chloroplast genome of _C. reinhardtii_, just beyond the end of the inverted repeat (Fig. 1), by hybridization with higher plant probes (Schmidt et al., 1985). If _sr-u-sm2_ proves to be a mutation in this protein, it will be the first case of protein-mediated streptomycin resistance to be identified in a chloroplast genome.

In summary, the ability to select mutants and to analyze them genetically makes _Chlamydomonas_ a superior model system for analysis of the biogenesis of chloroplast ribosomes and other components of the photosynthetic apparatus. We have identified more than twenty distinctive mutations affecting chloroplast ribosome assembly or function, and are actively pursuing characterization of these mutations at the molecular level.

This work was supported by NIH Grant GM-19427. We thank Mr. Bryce Burkhart for technical assistance, and Dr. Stephen Johnston for synthesis of oligonucleotide primers.

REFERENCES

Bartlett, S.G., et al. (1979) _Molec. Gen. Genet._ _176_, 199-208.
Bergmann, P., et al. (1985) _Plant Sci._ _39_, 133-140.
Bollen, A., et al. (1969) _Science_ _165_, 85-86.
Boynton, J.E., et al. (1972) _J. Cell Sci._ _10_, 267-305.
Boynton, J.E., et al. (1980) In: _The Harry Steenbock Symposium, Vol. 9, Ribosomes: Structure, Function and Genetics_, ed. G. Chambliss et al., University Park Press, Baltimore, pp. 903-950.
Brosius, J., et al. (1978) _Proc. Natl. Acad. Sci._ _75_, 4801-4805.
Chua, N.-H., N.W. Gillham (1977) _J. Cell Biol._ _74_, 441-452.
Davidson, J.N., et al. (1974) _Molec. Gen. Genet._ _132_, 119-129.
Davidson, J.N., et al. (1978) _Genetics_ _89_, 281-297.

Delepelaire, P. (1983) Photobiochem. Photobiophys. 6, 279-291.
Delepelaire, P. (1984) EMBO J. 3, 701-706.
Dron, M., et al. (1982a) J. Mol. Biol. 162, 775-793.
Dron, M., et al. (1982b) Nucl. Acids Res. 10, 7609-7620.
Erickson, J., et al. (1984) EMBO J. 3, 2753-2762.
Erickson, J., et al. (1985) In: Molecular Biology of the Photosynthetic Apparatus, ed. K. Steinback et al., Cold Spring Harbor, pp. 53-65.
Fromm, H., et al. (1985) Presentation at International Congress for Plant Molecular Biology, Savannah, GA.
Garnier, J., J. Maroc (1984) Adv. Photosyn. Res. 3, 183-186.
Gillham, N.W. (1978) Organelle Heredity, Raven Press, New York.
Gillham, N.W., et al. (1976) In: Genetics and Biogenesis of Chloroplasts and Mitochondria, ed. Th. Bücher et al., Elsevier-North Holland, Amsterdam, pp. 69-76.
Goldschmidt-Clermont, M. (1986) Plant Mol. Biol. 6, 13-21.
Goodenough, U.W., R.P. Levine (1970) J. Cell Biol. 50, 50-62.
Grant, D.M., et al. (1980) Proc. Natl. Acad. Sci. 77, 6067-6071.
Harris, E.H., et al. (1974) J. Cell Biol. 63, 160-179.
Harris, E.H., et al. (1976) In: Genetics of Algae, ed. R.A. Lewin, University of California Press, Berkeley, pp. 119-144.
Harris, E.H., et al. (1977) Molec. Gen. Genet. 155, 249-265.
Hildebrand, M., D. Bourque (1985) Presentation at International Congress for Plant Molecular Biology, Savannah, GA.
Huisman, J.G., et al. (1979) Planta 145, 351-356.
Malnoe, P., et al. (1979) J. Mol. Biol. 133, 417-434.
Marks, D.B., et al. (1986) Plant Physiol. 81, 702-704.
Merchant, S., L. Bogorad (1986) Mol. Cell Biol. 6, 462-469.
Mishkind, M.L., et al. (1985) J. Cell Biol. 100, 226-234.
Montandon, P.-E., et al. (1985) Nucl. Acids Res. 13, 4299-4310.
Myers, A.M., et al. (1984) Curr. Genet. 8, 369-378.
Ohta, N., et al. (1975) J. Biol. Chem. 250, 3655-3659.
Rochaix, J.-D. (1978) J. Mol. Biol. 126, 597-617.
Rochaix, J.-D., P. Malnoe (1978) Cell 15, 661-670.
Rochaix, J.-D., et al. (1984) Plant Molec. Biol. 3, 363-370.
Sager, R. (1972) Cytoplasmic Genes and Organelles, Academic Press, New York.
Salvucci, M.E., W.L. Ogren (1985) Planta 165, 340-347.
Sanger, F., et al. (1977) Proc. Natl. Acad. Sci. 74, 5463-5467.
Schlanger, G., R. Sager (1974) Proc. Natl. Acad. Sci. 71, 1715-1719.
Schmidt, R.J., et al. (1983) J. Cell Biol. 96, 1451-1563.
Schmidt, R.J., et al. (1985) In: Molecular Biology of the Photosynthetic Apparatus, ed. K. Steinback et al., Cold Spring Harbor, pp. 417-427.
Shepherd, H.S., et al. (1979) Proc. Natl. Acad. Sci. 76, 1353-1357.
Sigmund, C.D., et al. (1983) Nucl. Acids Res. 12, 4653-4663.
Sor, F., H. Fukuhara (1982) Nucl. Acids Res. 10, 6571-6577.
Sor, F., H. Fukuhara (1984) Nucl. Acids Res. 12, 8313-8318.
Spreitzer, R.J., W.L. Ogren (1985) In: Molecular Biology of the Photosynthetic Apparatus, ed. K. Steinback et al., Cold Spring Harbor, pp. 355-360.
Vallet, J.-M., J.-D. Rochaix (1985) Curr. Genet. 9, 321-324.
Waddell, J., et al. (1984) Nucl. Acids Res. 12, 3843-3856.
Watson, J.C., S.J. Surzycki (1982) Proc. Natl. Acad. Sci. 79, 2264-2267.
Woessner, J.P., et al. (1986) Gene 44, 17-28.
Woessner, J.P., et al. (1987) In preparation.
Wollman, F.-A., et al. (1980) J. Cell Biol. 87, 728-735.

EVIDENCE FOR MULTIPLE FUNCTIONS OF THE INTRINSIC, 32-34 KDA CHLOROPLAST MEMBRANE POLYPEPTIDE OF SCENEDESMUS IN PHOTOSYSTEM II REACTIONS.

NORMAN I. BISHOP
Department of Botany and Plant Pathology, Oregon State University
Corvallis, Oregon, 97331 U. S. A.

1. INTRODUCTION

A wealth of information has been developed in the past several years demonstrating that the reaction center core of photosystem II consists of minimally five major polypeptides with apparent molecular weights of 48, 44, 34, 32 and 10 kDa. These are recognized as the two apoproteins of two PS-II chlorophyll proteins, a protein of unknown, specific function, the Q_b-binding protein and cytochrome b-559 respectively. Chloroplast genes encoding for each of these components have been identified, sequenced and the amino acid sequence determined for each of these membrane components. The product of the psbA gene the 32-34 kDa intrinsic polypeptide of chloroplast membranes, has been equated with the rapidly metabolized chloroplast protein as originally discovered by Ellis and coworkers (1). It has also been described as the Q_b-binding protein, the D_1-protein and the herbicide-binding protein (2). The primary role of this protein has been deduced to be the site of binding of a plastoquinone molecule, Q_b, that functions in electron transport on the reducing side of PS-II; DCMU and atrazine compete at the Q_b binding site. This interpretation is supported by the preferential labeling of a 32-34 kDa PS-II core protein by the photoaffinity label, azidoatrazine (3). A comprehensive review of this subject area is presented in (2).

Through analysis of mutant phenotypes of Scenedesmus obliquus, blocked only in water photolysis (4) a 32-34 kDa polypeptide has been identified as an essential component of the oxidizing side of PS-II. Loss of this restricted portion of PS-II in the low fluorescent (LF) mutant, e.g., LF-1, is accompanied by the appearance of a 34-36 kDa polypeptide rather than the normal 32-34 kDa form (4). This approximate 2 kDa differential in apparent M_r is retained in PS-II reaction center core preparations of the WT and mutant LF-1 (5). Revertant strains of LF-1 possessing normal photosynthesis show recovery of the 32-34 kDa polypeptide (4). The observation by Metz, et al. that both the 32-34 and 34-36 kDa forms of this polypeptide are specifically labeled by azido[^{14}C]atrazine (5) questions the identification of the Q_b-protein as a moderator of electron transport only on the reducing side of PS-II. In this paper additional information gained from studies on the light sensitivity of growth, atrazine binding capacities and fluorescence properties of the LF-series of Scenedesmus mutants having either modified forms of or deficient in the 32-34 kDa polypeptide are presented. The data are interpreted to suggest that the rapidly labeled-rapidly metabolized 32-34 kDa chloroplast protein may function in reactions involving both the reducing and oxidizing sides of PS-II.

2. EXPERIMENTAL

For this study the wild-type and several non-photosynthetic mutant phenotypes of the green alga, Scenedesmus obliquus, were used as the experimental organisms. All methods for maintenance, growth, and mutant and revertant induction and recovery were according to procedures previously described (4). Algal chloroplast membranes used for this study were prepared from cells grown heterotrophically at 30C for 48 hrs. Procedures employed for rupturing of the algal cells, the buffers employed and centrifugation regimens were the same as previously described (4). Manganese, atrazine binding, and polypeptide analyses were performed on unfrozen chloroplast membrane fractions obtained by density gradient centrifugation (4).

Algal Development (Molecular and Cellular Aspects)
Ed. W. Wiessner, D.G. Robinson, and R.C. Starr

3. RESULTS AND DISCUSSION

To extend earlier efforts of this laboratory on the role of the intrinsic 32-34 kDa chloroplast membrane polypeptide in reactions of PS-II several new mutant phenotypes of Scenedesmus unable to synthesize normal levels of this protein have been isolated. Where appropriate, revertants of these new strains were also sought. In Tables 1 and 2 several parameters of whole cells and isolated chloroplast membranes of these new strains and their respective revertants are compared to those of the WT and mutant LF-1. The four mutant phenotypes display very low levels of photosynthesis but normal levels of photoreduction (data not shown) and of PS-I cell free reactions. Chloroplast reactions in which water serves as the electron donor are absent in all the mutants. The PS-II catalyzed reduction of dichlorophenol indophenol with diphenyl carbazide as the electron donor is retained near WT levels only in LF-1. Revertants developed from LF-1,. LF-18 and LF-22 show near normal levels of all reactions analyzed (Table I). Data of Table II illustrate that the loss of PS-II activity in mutants LF-18, LF-22 and LF-23 is accompanied by a decrease in chloroplast manganese and cytochrome $b\text{-}559_{HP}$ similar to that previously noted for LF-1 (4). The distinguishing features among the mutant strains are the noted inabilities to synthesize normal levels of the 32-34 kDa polypeptide and the apparently related photosensitivity of mixotrophic growth. Separation by LDS-PAGE and subsequent autoradiography of the polypeptides of $[^{35}]$S-labeled thylakoids of the mutant strains (Figs. 1 & 2) were used to estimate the amounts of the 32-34 kDa polypeptide shown in Table 2. Comparable analysis of the available revertant phenotypes (Table 2) confirmed the presence of this polypeptide in all cases.

TABLE 1. Comparison of various whole cell and isolated chloroplast particle photochemical reactions of the WT, select water-side mutant and respective revertants.

Strain	PS*	H20-DCPIP**	DPC-DCPIP**	DCPIP-MV
Wild type	1084	220	180	310
LF-1	120	0	168	386
LF-1 RVT-1	1016	206	186	325
LF-18	220	0	28	322
LF-18 RVT-1	1048	212	168	347
LF-22	116	0	25	316
LF-22 RVT	1050	220	175	366
LF-23	90	0	tr	320

*umoles O_2 produced/mg Chl-hr; **umoles DCPIP reduced/mg Chl-hr.

The observations that the 32-34 and 34-36 kDa polypeptides of WT and LF-1 are preferentially labeled by azidoatrazine (5), and, additionally, that mutant strains LF-18, LF-22 and LF-23 appear unable to synthesize the 32-34 polypeptide to varying degrees and show light-sensitive growth led us to evaluate the atrazine binding characteristics of these mutants. The results, as summarized in Figures 3 and 4 and Table 3, revealed that the atrazine binding constant (high affinity) and the number of binding sites are unaltered in LF-1 (and related phenotypes not considered here) but are acutely changed in the other mutants. As would be expected, the atrazine binding kinetics of the revertant phenotypes recovered to values comparable to those of WT chloroplasts.

TABLE 2. Manganese, cytochrome b-559(H.P.), relative abundance of "34 kDa" polypeptide and photoinhibition of heterotrophic growth in chloroplast particles and whole cells of WT, select water-side mutants and their respective revertants of Scenedesmus obliquus.

Strain Designation	Chl/Mn	Chl/cytochrome b-559 H.P.	"34 kDa" Protein*	Photoinhibition of growth**
Wild type	86 (4.6)	524	100	0
LF-1	302 (1.3)	2180	100***	20
LF-1RVT-1	111 (3.6)	590	100	0
LF-18	320 (1.2)	2200	25	68
LF-18RVT-1	81 (4.9)	560	100	0
LF-22	350 (1.1)		0	75
LF-22RVT-1	112 (3.6)	510	85	10
LF-23	360 (1.1)		0	75

Values are molecules chlorophyll/Mn atom (Mn atoms/400 chlorophyll molecules). Mole ratio of chlorophyll/cytochrome $b\text{-}559_{HP}$. *Estimated abundance (%WT) of the 32-34 kDa polypeptide. **Inhibition of growth (%WT) under mixotrophic growth conditions. Growth measured as chlorophyll content per 100 ul of cells. ***This values represents the level of the 34-36 kDa polypeptide in LF-1.

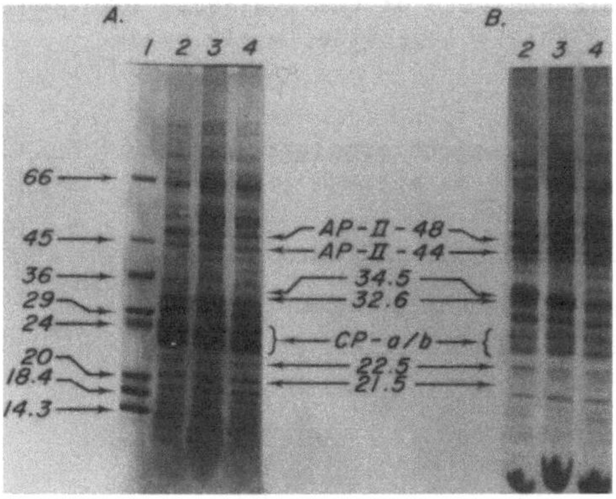

Figure 1. Analysis of purified, ^{35}S labeled chloroplast membrane polypeptides of WT, LF-1 and LF-18 by LDS-PAGE. (A) Slots #2, 3, and 4 contained samples equivalent to 10 ug protein of LF-1, WT and LF-18. Slot #1 contained protein standards of indicated MW. Coomassie Blue stained. (B). Autoradiographic pattern of gel slab illustrated in part A.

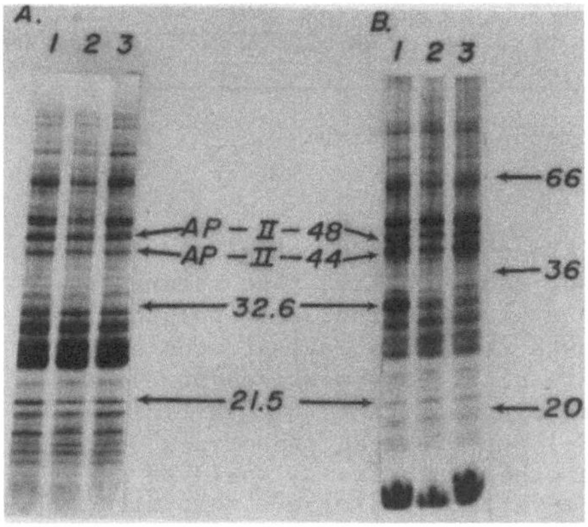

Figure 2. LDS-PAGE patterns of ^{35}S-labeled membrane polypeptides of WT and two low-fluorescent, water-side mutants, LF-22 and LF-23. (A.) Slots 1, 2, and 3 contained samples equivalent to 10 ug protein of WT, LF-2 and LF-23 chloroplast membranes. Pro- teins stained with Coomassie Blue. (B) Auto-radiographic pattern of gel slab shown in part A.

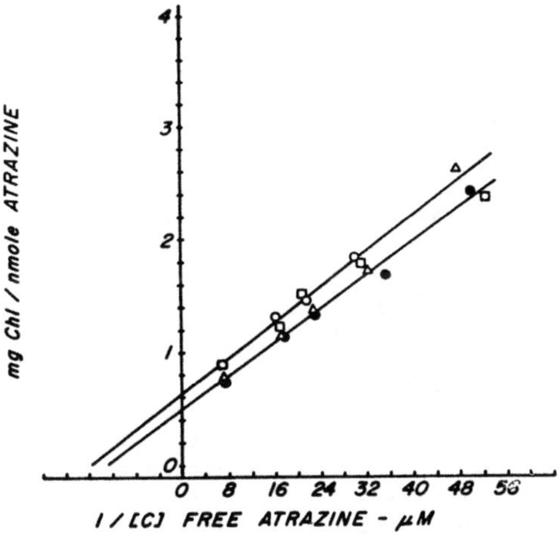

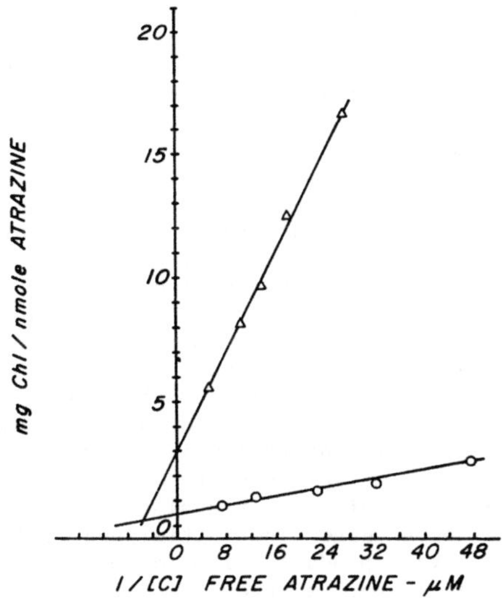

Figure 3 (left) Double reciprocal plots of [^{14}C]atrazine binding to chloroplast particles prepared from WT (●-●), LF-1 (○-○), LF-18RVT (△-△), and LF-1RVT. Values for K_b and the number of binding sites, n, (the mole ratio of bound atrazine/chlorophyll) were calculated from the slopes of the lines and of their intercepts with the abscissa, respectively. These values are presented in Table 3.

Figure 4 (right). Double reciprocal plots of [^{14}C]atrazine binding to chloroplast particles prepared from LF-18 (△-△) and LF-18RVT (○-○). Slope and intercept values presented in Table 3.

TABLE 3. Atrazine equilibrium binding constants and sites for chloroplast membranes of WT, mutants LF-1, LF-18, LF-22, LF-23 and revertants.

Strain	Binding Constant (K_b) (nanomolar)	# of Binding Sites (Chlorophyll/Atrazine)
WT	42	432
LF-1	43	468
LF-1 RVT-1	41	455
LF-18	165	2730
LF-18 RVT-1	40	417
LF-22	182	3235
LF-22 RVT-1	49	450
LF-23	168	3900

Based on the information presented here and the knowledge that antibodies prepared against the herbicide-binding protein of Chlamydomonas(6) and Anthirinum(7) show excellent cross-reactivity with the 32-34 kDa polypeptide of Scenedesmus chloroplasts , it is evident that this polypeptide is essential for the integrity of the oxygen evolving enzyme. Preferential genetic alteration or removal of it influence PS-II activities to varying degrees, including specific loss of water splitting activity accompanied by depletion of the normal manganese pool size, but lacking the photosensitivity observed in the strains with impaired abilities for the net synthesis of the 32-34 kDa polypeptide. Evidence that the inability of mutant strains, such as LF-23, to synthesize this polypeptide is accompanied by the loss of a specific chlorophyll-protein complex, i.e., the CP-43 noted by Machold (9), has not been found to date by detergent-gel electrophoresis. However, difference spectra of dark vs. light-treated cells of LF-23 suggest a rapid loss of a minor chlorophyll-bearing component upon exposure to light.

Based upon the observed primary sequence homology between the L and M subunits of the reaction center of Rhodopseudomonas viridis and the 32 kDa polypeptide (D_1-protein), Deisenhofer, et al (10) proposed that the D_1 and D_2 proteins constitute the core of the photosystem II reaction center. The lack of PS-II activity and the loss of the high affinity binding site in the mutant strains having reduced capability for net synthesis of the D_1-protein, even though possessing the 48 and 44 kDa polypeptides of the PS-II reaction center core, support this concept. Nothing can be specified from this study about the potential role of the D_2-polypeptide in PS-II of Scenedesmus since no clear resolution of it was obtained with the methodology employed. Metz, et al.(5) did note the presence of an intrinsic-type polypeptide in PS-II core preparations of WT and mutant LF-1 which they assumed to be the D_2 component.

The mutant phenotype represented by strain LF-1 perhaps is a consequence of the lack of post-translational processing of the primary translation product of the psbA gene. The initial nucleotide sequencing of this gene predicted a totally conserved product with a M_r of 38,950 (11) and several studies have shown evidence for a precursor form of the presumed D_1-protein. In LF-1, and related phenotypes, the unprocessed precursor is incorporated into the thylakoid such that PS-II activity develops, i.e., a DCMU-sensitive electron transport from diphenylcarbazide to dichlorophenol indophenol and presence of a variable-yield fluorescence, but without the development of PS-II photosensitivity. We suggest that the inability of this phenotype to perform water photolysis results from the presence of the lengthened carboxy-terminal end of the D_1-protein in the thylakoid lumen. Specifically it is suggested that specific interactions among the polypeptide strands connecting the membrane spanning helices (10,12) and the nuclear encoded, extrinsic PS-II polypeptides, cytochome b-559 and chloroplast-manganese would be affected. Antisera prepared against decapeptides synthesized according to the sequence information of the hydrophyllic, surface exposed regions predicted by models of the D_1-protein (12) have been utilized by Sayre, et al. (13) to examine the secondary and tertiary structure of this protein. Noteworthy for our studies was the observation that the antisera prepared against two decapeptide sequences (#2 and #3) of the amino acid strand connecting the A and B helices were bound effectively to thylakoids only if the extrinsic, PS-II polypeptides had been removed. (Sayre, personal communication). This lumenal component also contains an helical conformation (amino acids 74-96) which may serve in the manganese liganding between D_1 and the extrinsic 33 kDa polypeptide of Murata. This suggested interaction might also be altered by the presence of the unprocessed, carboxy-terminus strand. Recovery of the post-translational clipping step, as noted in the revertant phenotype of LF-1, would result in "normal" PS-II activity comparable to that of the wild type Scenedesmus.

ACKNOWLEDGMENTS

The support for the research reported in these pages by the United States Department of Agricultural (82-CRCR-1-1079) is gratefully recognized.

REFERENCES

1 Ellis, J.R. (1981) Ann. Rev. Plant Physiol. 32, 111-137.
2 Kyle, D.J. (1985) Photochem. Photobiol. 41, 107-116.
3 Pfister, K., Steinbach, K. E., Gardner, G. and Arntzen, C. J. (1981) Proc. Natl. Acad. Sci. USA 78, 981-985.
4 Bishop, N.I. (1984) In: Advances in Photosynthesis Research (Sybesma, C.,ed.) M. Nijhoff/Dr W.Junk Publ. The Hague. I, pp.321-332.
5 Metz, J.B., Bricker, T.M. and Seibert, M. (1985) FEBS Lett., 185, 191-196.
6 Wildner, G., Berzborn, R., Dedner, N., Dmoch, R. and Fiebig, C. Proc. VII International Congress oon Photosynthesis. (In Press).
7 Metz, J.B., Pakrasi, H.B., Seibert, M. and Arntzen, C. J. FEBS Lett.,In Press).
8 Kyle, D.J., Ohad, I. and Arntzen, C.J. (1984) Proc. Natl. Acad. Sci. USA 81, 4070-4074.
9 Machold, O. Proc. VII International Congress on Photosynthesis. (In Press).
10 Deisenhofer, J., Epp, O., Miki, K., Huber, R. and Michel, H. (1985) Nature 318: 618-624.
11 Zurawski, G., Bohnert, H. J., Whitfeld, P. R. and Bottomley, W. (1982) Proc. Natl. Acad. Sci. USA 79: 7699-7703.
12 Trebst,A. (1985) Z. Naturforsch. 41c: 240-245.
13 Sayre, R. T., Andersson, B. and Bogorad, L. Proc. VII International Congress on Photosynthesis. (In Press).

"LIFE CYCLE" OF THE 32 kDa-PROTEIN OF CHLAMYDOMONAS CHLOROPLASTS

Ralf Hartmann and Michael Wettern
Botanisches Institut der Technischen Universität Braunschweig
Humboldtstraße 1, D-3300 Braunschweig, F.R. Germany

Abstract

The turnover of the DCMU- and atrazine-binding 32 kDa-polypeptide of photosynthetic membranes has been measured by radioactive pulse labelling, chase experiments, double labelling, and photoaffinity labelling. The synthesis of this protein is stimulated by light and takes place at thylakoid-bound ribosomes of unstacked thylakoid membranes. The doubling-time in synthesis was about 40 min. The pool size of this "transient pool" of the 32 kDa-molecules, which are not yet functionally integrated, was calculated as 30-35 % of total 32 kDa-polypeptides. These molecules do not bind the herbicide atrazine. Under the experimental conditions used the transfer from unstacked to stacked membranes takes about 4 hrs. In the stacked membranes the 32 kDa-polypeptides are functionally integrated within active PS II complexes. Only these integrated molecules bind atrazine and are subjected to light stimulated degradation. Under light stress conditions at 2000 $\mu E/m^2/sec$ for 90 min the oxygen evolution declines and the 32 kDa-polypeptide is preferentially synthesized <u>and</u> degraded at the same time (besides polypeptides of the RC II and LHC). During these stress conditions the specific radioactive labelling of fatty acids (16:0, $16{:}1_{tr}$ and 18:2) is strongly diminished but restored in a following recovery. A thylakoid-bound thiolendoprotease is involved in the degradation process of the 32 kDa-polypeptide. Only under <u>in vitro</u> conditions, but not <u>in vivo</u>, cleaved breakdown products (with M_r larger than 10 kDa) are released from the membranes and accumulated. A phosphorylation step is involved in the <u>in vitro</u> digestion of the polypeptides. A scheme of the "life cycle" of the herbicide-binding 32 kDa-polypeptide from thylakoid membranes will be given.

Algal Development (Molecular and Cellular Aspects)
Ed. W. Wiessner, D.G. Robinson, and R.C. Starr

Introduction

Although the 32 kDa-polypeptide (D-1) is a minor component of thylakoids in algae and higher plants, much attention has been focused on it. It was shown to be the binding site for herbicides (Pfister et al. 1981), the rapidly synthesized polypeptide of thylakoid membranes (Edelman and Reisfeld 1978; Owens et al. 1982) and to participate in Q_B binding (Arntzen et al. 1984; Kyle et al. 1984; Ohad et al. 1984). Measurements of the stoichiometry of the various PS II polypeptides in C. reinhardii showed that thylakoids contain about one herbicide binding polypeptide per one PS II unit and the accelerated light stimulated turnover of the 32 kDa-polypeptide was postulated to be the consequence of its function in quinone reduction. Based on comparison of amino acid sequence and hydropathy profile homologies of D-1 and D-2 with the subunits of the bacterial photosystem, recently a model was presented, in which the reaction center of photosystem II in algae and higher plants is formed by the polypeptides D-1 and D-2 (Trebst 1986). The structural gene of the 32 kDa-polypeptide (psbA) has been mapped on the chloroplast DNA and cloned from various plant species (for a review: Kyle 1985). The polypeptide is synthesized as a precursor on ribosomes bound to unstacked lamellae and there processed to its mature form (Edelman and Reisfeld 1978; Herrin and Michaels 1985). In C. reinhardii the D-1 polypeptide is phosphorylated by a light-activated membrane bound kinase, which might also phosphorylate the LHC polypeptides, and dephosphorylation of the D-1 polypeptide alters its herbicide binding affinity (Owens and Ohad 1982).

In our studies on the regulation of the turnover of the 32 kDa-polypeptide in C. reinhardii (Wettern 1984; Wettern and Ohad 1984) and Chlorella saccharophila (Gerloff and Wettern 1986) we observed a lag phase of about 4 hrs before a decrease of radioactivity incorporated into the 32 kDa-polypeptide occurred. In the present work evidence will be reported for the existence of two pools of the 32 kDa-polypeptide and data on its turnover during light stress and following recovery will be presented. Changes of fatty acid synthesis during light stress and recovery are discussed in respect to their possible role for the functional integration and degradation of the 32 kDa-polypeptide.

Results and Discussion

The herbicide-binding protein is part of the PS II units, which are located in the stacked fractions of thylakoid membranes. As in Spirodela oligorrhiza (Mattoo et al. 1981), several higher plants (Hoffman-Falk et al. 1983) and C. reinhardii (Wettern et al. 1983), it is probably the only thylakoid membrane polypeptide showing a high turnover rate. This behaviour was used for its identification in LDS-PAGE. To analyse the observed lag phase in loss of radioactivity from the

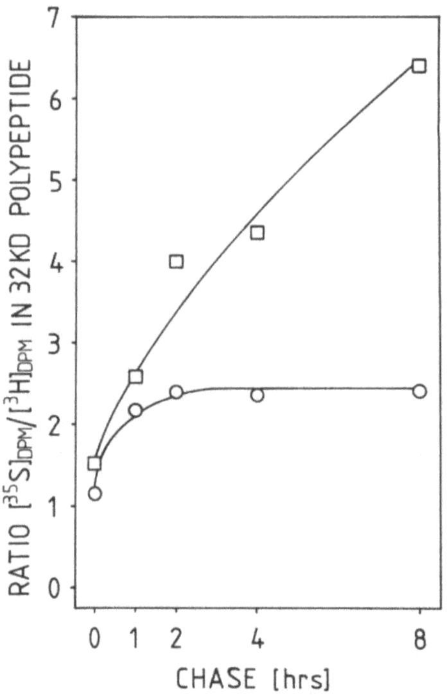

Figure 1

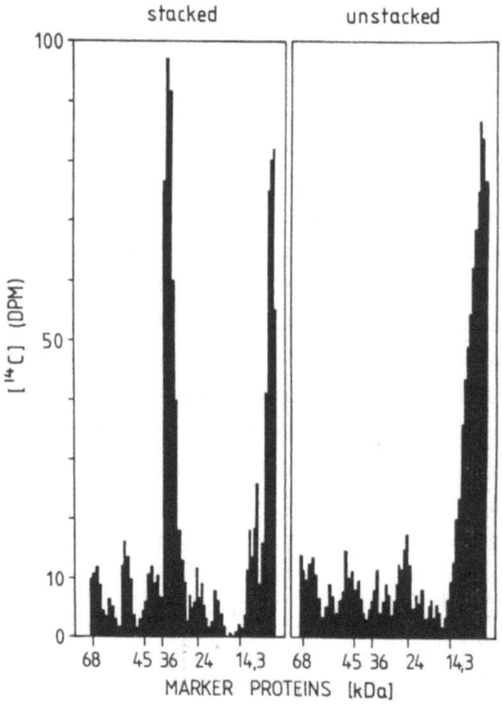

Figure 2

Figure 1. Ratios of [^{35}S]/[^{3}H] in the 32 kDa-polypeptide from stacked (□) and unstacked (○) thylakoid membranes. Cells of C. reinhardii were first labelled with [^{3}H]acetate, washed with nonradioactive medium and then labelled with [^{35}S]sulphate. Each labelling lasted 30 min. During a chase experiment, samples were taken as indicated, membranes fractionated by digitonin treatment and the fractions isolated, the polypeptides separated by LDS-PAGE, the bands of the 32 kDa-polypeptides sliced, dissolved, the radioactivity measured and the ratios calculated.

Figure 2. Photoaffinity labelling of C. reinhardii cells with [^{14}C]azido-atrazine. After photoaffinity labelling, polypeptides of the membrane fractions were prepared as in Fig. 1, the gel sliced, the strips dissolved with H_2O_2 and radioactivity measured.

32 kDa-polypeptide (Wettern 1984) a double-labelling experiment with successive labelling was performed. During a chase samples were taken at various times and prepared as described (Fig.1). The observed constant value of ^{35}S over ^{3}H after 2 h in unstacked membranes and the increase of this ratio during the entire chase experiment in the stacked membrane fractions is explained by a transport of newly synthesized 32 kDa-polypeptides from unstacked thylakoids to stacked membranes. A degradation of the polypeptide in both membrane fractions would have resulted (assuming a random degradation in both fractions) in a constant ratio in each of the fractions. The pool size of the non-functionally integrated 32 kDa-polypeptides in unstacked thylakoids was calculated as 30-35 % of total 32 kDa-polypeptides in the membrane. Support for the model of different sites for synthesis and degradation of the 32 kDa-polypeptide is given by the observation that at least in <u>C. reinhardii</u> this polypeptide is synthesized on thylakoid-bound ribosomes (Herrin and Michaels 1985) and that in this alga the surface of unstacked thylakoids contributes more to the total surface of thylakoids than the stacked fractions do. Figure 2 shows, that only those molecules of the 32 kDa-polypeptides, which are functionally integrated in PS II units in the stacked thylakoid membranes, bind to azido-atrazine. For binding, azido-atrazine needs its respective binding protein in an appropriate environment (Wettern 1986). The differences in the environment for the two pools of the 32 kDa-polypeptides we do not know yet. It might be due to different protein-protein or protein-lipid interactions in the two thylakoid membrane fractions. Moreover, the 32 kDa-polypeptide might undergo posttranslational modification (acylation, glycosylation ?), necessary for its proper integration, and thereby allowing binding of hercides. As it will be shown later, fatty acids of thylakoid membranes might play a key role for this integration. The importance of the membrane lipid phase for specific herbicide binding was demonstrated by the fact, that this binding should occur via an equilibrium of herbicides solubilized in the lipid phase of membranes rather than via those herbicide molecules solubilized in the stroma of chloroplasts (Laasch 1984). A quite novel posttranslational modification associated with certain chloroplast proteins including the 32 kDa-polypeptide was presented recently (Edelman and Mattoo 1985).

The photosynthetic capacity, turnover of thylakoid membrane polypeptides, and rates of synthesis of fatty acids during high light intensities and a following recovery period were studied in C. reinhardii as tools for a better understanding of the regulation of synthesis and degradation of the herbicide binding 32 kDa-polypeptide. White light of 3000 µE/m²/sec at 25 °C completely destroys the photosynthetic capacity of the algal cells (Fig. 3). The photosynthetic rate declines and the cells are not able to recover from the exposure. This results in a complete inability to synthesize the 32 kDa-polypeptide during the recovery period (data not shown). Exposing the algae to light of 2000 µE/m²/sec for 90 min results in a decrease of oxygen evolution which is restored during the following recovery period. During this high light treatment, the 32 kDa-polypeptide is synthesi-

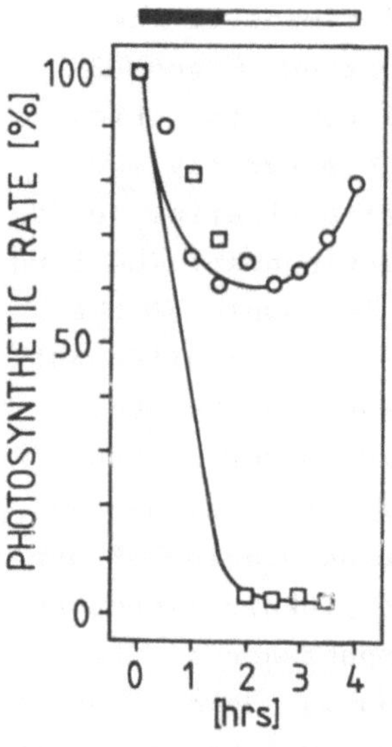

Figure 3

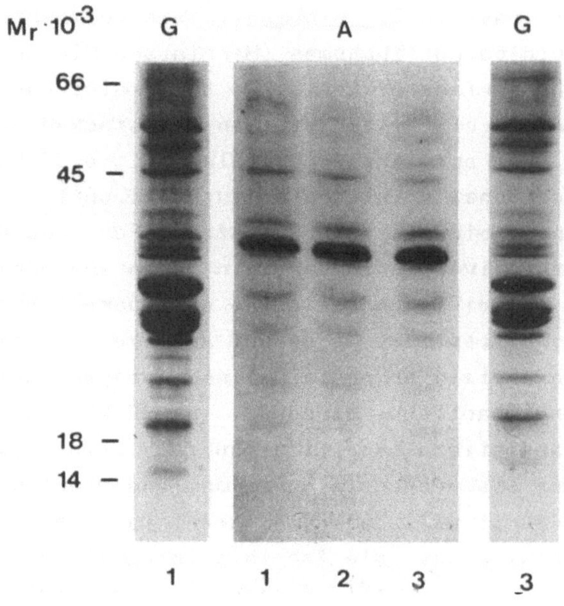

Figure 4

Figure 3. Oxygen evolution from C. reinhardii cells during photoinhibition and a following recovery period. Algal cells were concentrated to a cell number of $2 \cdot 10^7$ cells per ml and kept at 25 °C during photoinhibition at 2000 µE/m²/sec (○) or at 3000 µE/m²/sec (□). Light intensities during recovery periods were 1/10 of that used for photoinhibition. 100 % photosynthetic rate = 47.9 µmol O_2 per mg chlorophyll and hour. ▬: photoinhibition; ▭: recovery.

Figure 4. Pulse labelling of C. reinhardii during photoinhibition at 2000 µE/m²/sec with [^{35}S]sulphate. Each labelling lasted 30 min, samples were prepared as in Fig. 1 and autoradiographs developed as described (Wettern and Galling 1985). Lane 1-3: labelling between 0-30, 30-60, and 60-90 min, respectively, of photoinhibition. G: staining for protein, A: autoradiograph.

zed very fast (Fig. 4) compared with its rate of synthesis at normal light intensities (Wettern and Ohad 1984). Although the photosynthetic oxygen evolution declines, the polypeptide still shows a turnover (Fig. 5). It is synthesized and degraded at the same time during the light stress condition. Under the chosen conditions of light stress and labelling time, not only the 32 kDa-polypeptide but also proteins in the molecular weight range of 45 and 28 kDa are damaged. With the data obtained, we are at present not able to decide, whether the rate of synthesis of the 32 kDa-polypeptide equals its rate of degradation (Fig.5) or not. Therefore, we do not know if the turnover is still in

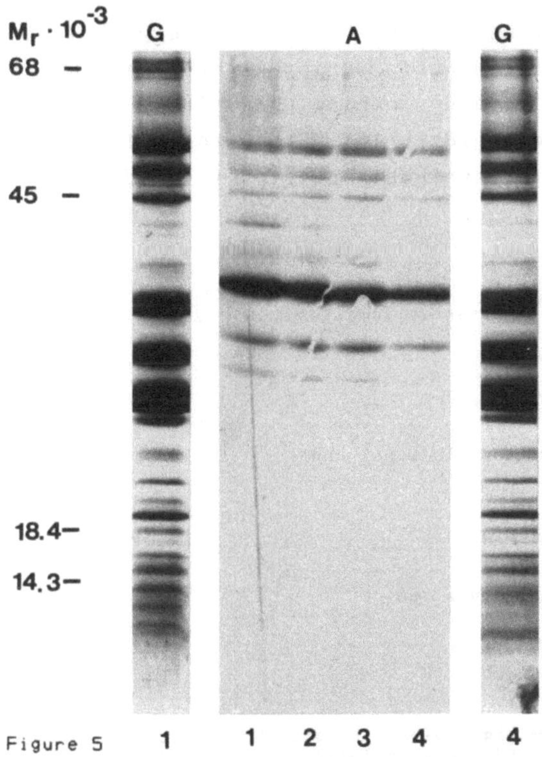

Figure 5

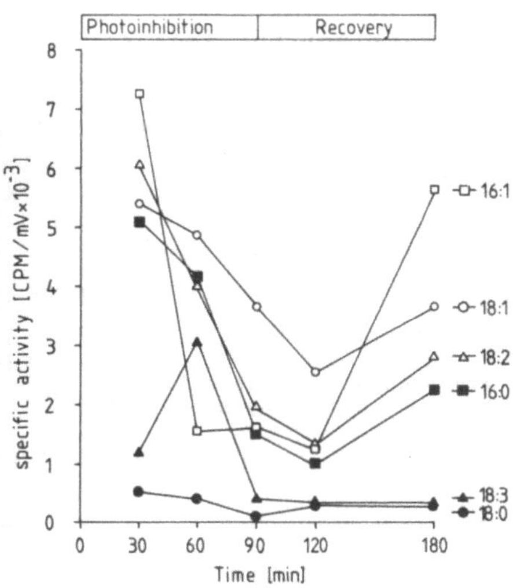

Figure 6

Figure 5. Loss of radioactivity during photoinhibition at 2000 µE/m²/sec of ^{35}S pulse labelled cells of C. reinhardii. Cells were labelled with [^{35}S]sulphate for 30 min prior to the photoinhibition, washed and resuspended in fresh growth medium. Samples were taken at times indicated and prepared as in Fig. 4. Lane 1-4: Samples taken at the end of pulse and after 30, 60, and 90 min, respectively, during photoinhibition. G: staining for protein, A: autoradiograph.

Figure 6. Changes in the rates of synthesis (specific activities) of fatty acids in cells of C. reinhardii during photoinhibition at 2000 µE/m²/sec and a following recovery period at 200 µE/m²/sec. During photoinhibition and recovery samples of algae were labelled with [^{14}C]acetate. Each labelling lasted 30 min. After times indicated, total cell fatty acids were extracted, fatty acid pattern and the radioactivity in each fatty acid measured. Data are given as the ratio of radioactivity in each fatty acid over the content of the same fatty acid.

balance during the high light exposure to permit the algal cells to survive. The synthesis of various fatty acids in C. reinhardii cells is strongly affected during condition of light stress (Fig. 6). The strongest decline in the rate of synthesis is exhibited by the trans-3-hexadecanoic acid ($16{:}1_{tr}$) which is exclusively esterified at position C-2 of the glycerol backbone in phosphatidylglycerol. The fatty acid $16{:}1_{tr}$, but not the α-linolenic acid (α-18:3), was shown to be essential for the efficient photofunction of thylakoids (Leech and Walton 1983). Although $16{:}1_{tr}$ stabilize the LHC protein in vitro, its function is more subtle than previously proposed for grane stacking (review: Frentzen 1986). Whatever the correct function of this fatty acid in thylakoid membranes might be, it is the fatty acid with most pronounced decline in the rate of synthesis during light stress which is restored during the following recovery period. The effect on the rate of synthesis of the fatty acids 18:1, 18:2, 16:0 and α-18:3 is still distinct but less strong. The synthesis of stearic acid (18:0)

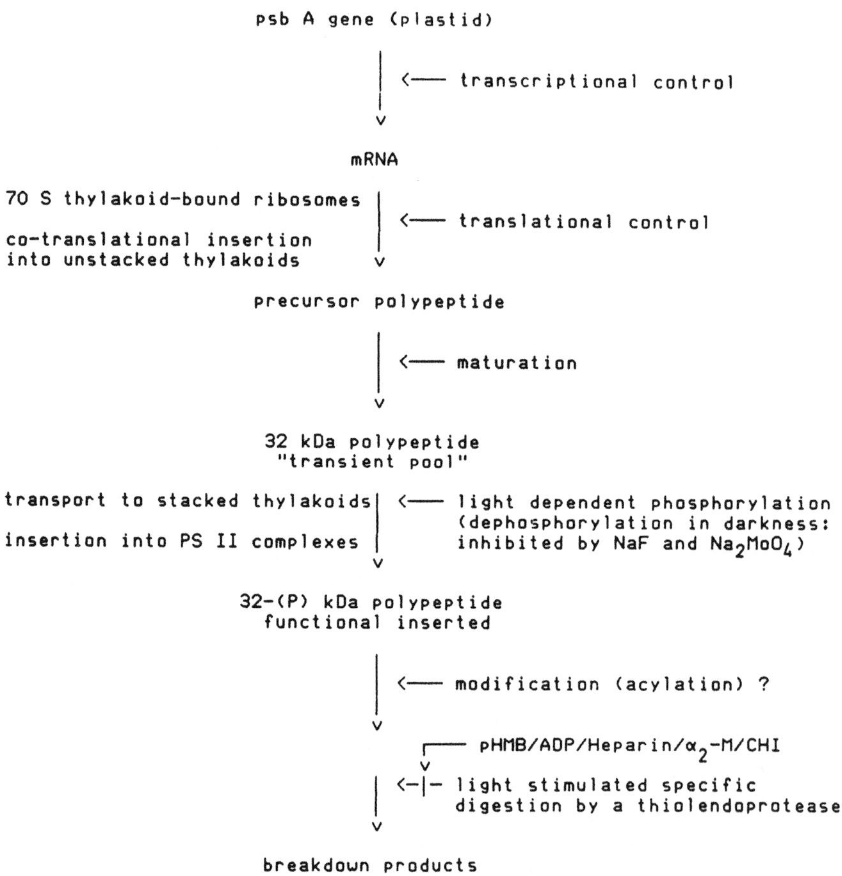

Figure 7. "Life cycle" of the herbicide binding 32 kDa-polypeptide For details see text.

seems not to be affected at all during high light condition and the following recovery period. At present we do not know the role of fatty acids in thylakoid membranes, but they might play an intriguing role in the stabilization of thylakoid membrane polypeptides.

In Fig. 7 we summarized the "life cycle" of the herbicide-binding 32 kDa-polypeptide. It is coded for by the chloroplast gene psbA and synthesized as a precursor by ribosomes bound to unstacked membranes. After maturation it is transported from a "transient pool" to stacked membranes and inserted into PS II units when a site becomes available. The protein is phosphorylated and perhaps additionally modified. Dependent on the light intensity it is degraded by a thylakoid-bound thiolendoprotease in an ATP-consuming reaction (Wettern and Galling 1985). Breakdown products accumulate only in an in vitro system.

Acknowledgements: Thanks to Dr. Galling for discussion, Dr. Oettmeier providing labelled azido-atrazine, T. Teucher performing fatty acid analysis, and U. Nieländer for excellent technical assistance.

References

Arntzen, C.J., D.J. Kyle, M. Wettern, I. Ohad (1984) In: Biosynthesis of the Photosynthetic Apparatus: Molecular Biology, Development and Regulation. ed. J.P. Thornber, L.A. Staehelin, R.B. Hallick, Liss, New York, pp. 313-324.
Edelman, M., A. Reisfeld (1978) In: Chloroplast Development. ed. G. Akoyunoglou, J.H. Argyroudi-Akoyunoglou, Elsevier, Amsterdam, pp.641-652.
Edelman, M., A.K. Mattoo (1985) In: First Internat. Congr. of Plant Molec. Biology. ed. G.A. Galau, Abstracts, pp. 54.
Frentzen, M. (1986) J. Plant Physiol. 124, 193-209.
Gerloff, P., M. Wettern (1986) In: Regulation of Chloroplast Differentiation. ed. G. Akoyunoglou, H. Senger, Liss, New York, pp. 187-192.
Herrin, D., A. Michaels (1985) FEBS Lett. 184, 90-95.
Hoffman-Falk, H., A.K. Mattoo, J.B. Marder, M. Edelman, R.J. Ellis (1983) J. Biol. Chem. 257, 4583-4587.
Kyle, D.J., I. Ohad, R. Guy, C.J. Arntzen (1984) In: Advances in Photosynthesis Res. Vol III. ed. C. Sybesma, Nijhoff and Jungk, The Hague, pp. 67-70.
Kyle, D.J. (1985) Photochem. and Photobiol. 41, 107-116.
Laasch, H. (1984) Ph.D. Dissertation, Würzburg University, F.R. Germany.
Leech, R.M., C.A. Walton (1983)In: Biosynthesis and Function of Plant Lipids. ed. W.W. Thomson, J.B. Mudd, M. Gibbs, Am Soc Plant Physiol, Rockville, pp. 81-103.
Mattoo, A.K., U. Pick, H. Hoffman-Falk, M. Edelman (1981) PNAS 78, 1572-1576.
Ohad, I., D.J. Kyle, C.J. Arntzen (1984) J. Cell Biol. 99, 481-485.
Owens, G.C., I. Ohad (1982) J. Cell Biol. 93, 712-718.
Owens, G.C., M. Wettern, N. Lavintman, D. Ish-Shalom, G. Schuster, D. Kirilovsky, I. Ohad (1982) In: Cell Function and Differentiation, Part B. ed. G. Akoyunoglou et al. Liss, New York, pp. 223-231.
Pfister, K., K.E. Steinback, C.J. Arntzen (1981) In: Photosynthesis, VI. ed. G. Akoyunoglou, Balaban, Philadelphia, pp. 595-606.
Trebst, A. (1986) Z. Naturforsch. 41c, 240-245.
Wettern, M., G.C. Owens, I. Ohad (1983) Methods Enzymol. 97, 554-567.
Wettern, M. (1984) In: Advances in Photosynthesis Res. III. ed. C. Sybesma, Nijhoff and Jungk, The Hague, pp. 179-182.
Wettern, M., I. Ohad (1984) Israel J. Bot. 33, 253-263.
Wettern, M., G. Galling (1985) Planta 166, 474-482.
Wettern, M. (1986) Plant Science 43, 173-177.

TWO DIFFERENT FORMS OF LHC IN THE GREEN ALGA *CHLAMYDOBOTRYS STELLATA*

B.Kohnke, P.Brandt, D.Mende and W.Wiessner
Abt. Experimentelle Phykologie, Pflanzenphysiologisches Institut, Universität Göttingen, Untere Karspüle 2, D-3400 Göttingen, F.R.G.

K. Kloppstech
Institut für Botanik, Universität Hannover, Herrenhäuser Str. 2
D-3000 Hannover, F.R.G.

A. Radunz
Lehrstuhl für Zellphysiologie, Universität Bielefeld
D-4800 Bielefeld 1, F.R.G.

Photosynthetic CO_2 fixation requires the cooperation of two separate photosystems, PS I and PS II, operating in series. These two photosystems are composed of several well defined membrane protein complexes. PS I contains the reaction center chlorophyll a-protein (P680) and associated antenna chlorophyll proteins. PS II is an association between the reaction center chlorophyll a-protein (P700) and bound antenna pigments. P680 and P700 are directly engaged in charge separation across the thylakoid membrane. The antenna pigments absorb light and channel the excitation energy to the reaction centers. About 50 % of the total chlorophyll content of the thylakoids is organized in these two complexes. The remaining chlorophyll a and the chlorophyll b (in case of chlorophyll b containing eukaryotes) belong to the light-harvesting chlorophyll a/b protein complex, LHC II. It also absorbs photosynthetically usable light and was originally believed to interact preferentially with PS II. Evidence is now accumulating that phosphorylation of LHC can increase transfer of energy to PS I (α-transfer; Bennett et al. 1980, review Bennett 1983) increasing energy distribution to PS I at the expense of PS II. Thus phosphorylation could play a major role in distributing excitation energy between the two photosystems.

The green alga *Chlamydobotrys stellata* grows autotrophically with carbon dioxide as carbon source but also by a light-driven assimilation of acetate, e.g. photoheterotrophically, (photoorganotrophically, Pringsheim and Wiessner 1960, Gaffron et al. 1963). The latter needs only the operation of PS I for cyclic photophosphorylation (Wiessner 1965). This makes necessary a preferential channelling of excitation energy to PS I and a decrease of energy flux

Algal Development (Molecular and Cellular Aspects)
Ed. W. Wiessner, D.G. Robinson, and R.C. Starr

to PS II. It has been the aim of the work reported here to find out how this preferential supply of energy to PS I is brought about in *Chlamydobotrys stellata*. We present evidence on the existence of a special PS I connected light-harvesting chlorophyll a-protein complex (LHCa, formerly called LHCPa) in acetate assimilating *Chlamydobotrys stellata*, the biogenesis of which is regulated by the presence of acetate.

Material and Methods

Growth of Cells: *Chlamydobotrys stellata*, strain 10-1e (Sammlung von Algenkulturen, Pflanzenphysiologisches Institut, Universität Göttingen, F.R.G.) was cultivated in continuous light either autotrophically with CO_2 or photoheterotrophically with acetate as carbon source (Wiessner 1968). *Chlorella fusca* (strain 211-8b) and *Chlamydomonas reinhardii* (strain 11-32b) were cultivated in the growth medium of Kuhl (1962) either autotrophically or photoheterotrophically with acetate.

Measurement of Fluorescence Emission : 683 nm fluorescence at 298 K was measured as described by Mende and Wiessner (1985). For experiments on state 2 - state 1 transition the algae were preilluminated for 15 min either at 620 nm (light 2) or at 715 nm (light 1). Fluorescence at 77 K was recorded using the instrumental setup of Harnischfeger and Codd (1977, 1978). Excitation light was obtained by a broad band blue light interference filter (BALZERS). The algae were kept in darkness from the time of their harvesting from the cultures until their exposure to the excitation light. Fluorescence inductions at 693 and 718 nm were recorded simultaneously.

Protein synthesis :Light dependent protein synthesis was determined by ^{14}C incorporation into proteins from ^{14}C-sodium bicarbonate in case of autotrophic cells and from ^{14}C-sodium acetate (AMERSHAM) when the algae were grown photoheterotrophically (Brandt and Wiessner 1984).

Isolation of Thylakoids, Separation of Plastid Membrane Proteins : For preparation of thylakoids, cells were collected during their exponential growth phase and broken in a YEDA press (Res. and Dev.

Co. Ltd. Rehovot, Israel). Thylakoid fragments were isolated as described by Brandt (1980). Separation of plastid membrane proteins was done without heat denaturation in 15 % acrylamide gels containing 1 % Deriphat (Markwell et al. 1979). After Coomassie blue staining,the gels were treated for fluorometry as described by Bonner and Lakey (1974).

Peptide Mapping of LHC polypeptides was done according to Cleveland et al. (1977) with Staphylococcus aureus V-8 protease (SIGMA). The electrophoretic analysis of the products of proteolysis occurred as described below.

Western Blotting and Immunodetection of LHC polypeptides were carried out according to Towbin et al. (1979). After isolation of thylakoids and solubilization, the structural proteins were separated on one-dimensional polyacrylamide gels and transferred onto nitrocellulose sheets. These were cut into 5 mm wide strips and incubated with antibodies against LHC. The immunological reaction was made visible with peroxidase-conjugated antibody (SIGMA).

Extraction of m-RNA : 6 g *Chlamydobotrys stellata* (1g algae/10 ml buffer) were frozen to liquid nitrogen and poly(A^+)-containing m-RNA was prepared as described by Apel and Kloppstech (1978).

Wheat-germ Cell-free Translation : The wheat-germ S30-fraction was prepared according to Roberts and Paterson (1973). The reaction mixture contained in a final 40 µl volume 10 µl wheat-germ (130 OD, 280 nm) and the following constituents : 100 mM $Mg(CH_3COO)_2$; 0,1 mM spermine; 25 µM each of the 19 amino acids (except methionine); 1 mM ATP; 25 µM GTP; 25 µM CTP; 8 mM creatine phosphate; 1 µg creatine-phosphokinase; 20 mM HEPES/KOH pH 7,1; 2 mM DTE; 4 µg wheat-germ t-RNA and 12 µCi ^{35}S-methionine (AMERSHAM). The samples were incubated for 90 min at 25° C. The radioactivity of the hot TCA insoluble fraction was estimated in 5 µl aliquots by the filter disc method of Mans and Novelli (1961). The remainder was either precipitated with 5 % TCA and prepared for electrophoresis or used for immunoprecipitation with antibody raised against LHC.

Preparation of Antisera to LHC II and LHCa occurred by injection of electrophoretically purified antigen-proteins into rabbits as described by Radunz (1975). 1 mg antigene denaturated in 1 % natrium-dodecylsulfate were suspended in 1 ml 0,06 M phosphate buffer pH 7,4, and emulsified with 1 ml of Freund's adjuvant. This emulsion was subcutaneously injected into the back of a rabbit. After 30 days with a five day interval two intravenous injections of each time 1 mg protein in the aforementioned buffer were made. The first blood withdrawal was made 10 days after the last injection. The isolated antibodies against LHCa and LHC II were shown to be monospecific by double diffusion assays and cross immunoelectrophoresis.

Immunoprecipitation of in vitro Translation Products : The translation mixture was diluted five fold with PBS buffer (0.15 M NaCl; 0.1 % SDS; 1 % Triton x-100) prior to the addition of 2.5 µl of antibody to LHC per 75 µl of the translation supernatent. After an incubation for 12 h at 4° C, the immunoglobins were removed by adsorption to 50 µl of a suspension of *Staphylococcus aureus* membranes (Kessler 1976). After 30 min the precipitated immunoglobin-protein complexes were washed three times with PBS and eluted at 100° C as described by Neville (1971) prior to further electrophoretic analysis.

Electrophoretic Analysis of Translation Products : The polypeptides were solubilized in the presence of sodium dodecylsulfate and separated according to Laemmli (1970). Coomassie blue staining and treatment for fluorography according to Bonner and Laskey (1974).

Results and Discussion

The role of LHC phosphorylation in excitation energy transfer to PS I in *Chlamydobotrys stellata* was tested by experiments on state 2 - state 1 transition. The protein kinase involved in this phosphorylation is responsive to the redox state of the plastoquinone. It is activated by a reduced, and inhibited by an oxidized plastoquinone pool. Reduction is facilitated by light 2, e.g. light preferentially absorbed by PS II (state 2). By contrast, a preferential activation of PS I by light 1 oxidizes the pool (state 1; Allen and Horton 1981, Allen et al. 1981; Bennett 1979; Horton et al. 1981; Steinback et al.

1982). Light 2 therefore decreases room temperature PS II-fluorescence because of kinase activation, and comparative measurements of fluorescence after preillumination with light 2 respectively light 1 can reveal the ability of a given photosynthetic organism to regulate energy distribution to PS I at the level of LHC phosphorylation. With *Chlorella fusca* light 2 decreases F-683 intensity (Fig.1). However, the application of light 1 or light 2 does not affect fluorescence intensity of autotrophically or photoheterotrophically cultivated *Chlamydobotrys stellata*. Obviously in *Chlorella*, but not in *Chlamydobotrys*, the energy flux to PS I (α-transfer) is regulated by reversible LHC-phosphorylation.

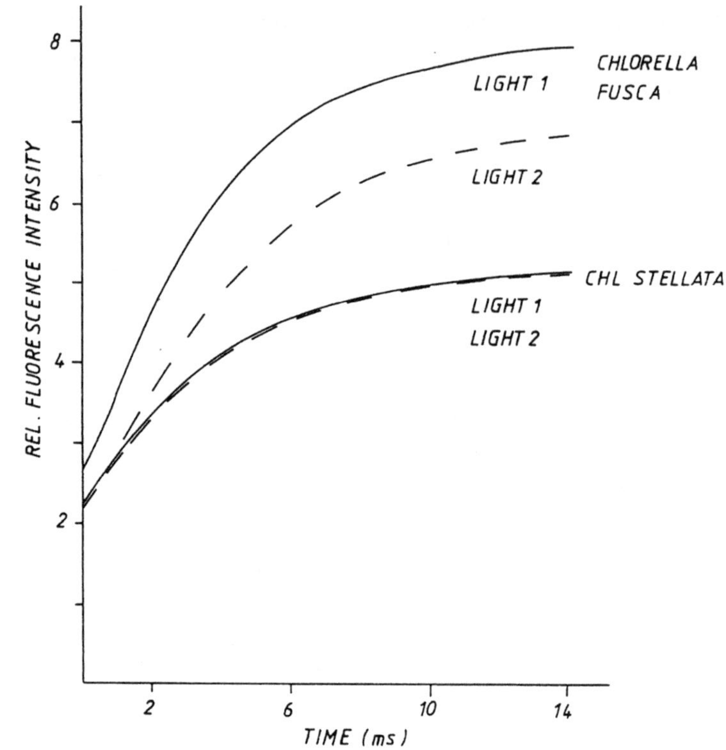

Fig. 1 : 683 nm fluorescence induction at room temperature of *Chlorella fusca* and *Chlamydobotrys stellata* after preillumination with light 1 (-----), respectively with light 2 (- - -)

A second way to influence energy distribution between the two photosystems is the direct transfer of excitation energy from PS II to PS I (spillover). This was checked by measurements of PS I and PS II fluorescence at 77 K when only primary photochemistry is possible. As has been pointed out by Butler and Strasser (1978), the kinetics of fluorescence induction yield changes are theoretically identical for PS II- (F-693) and PS I-fluorescence (F-718) under this condition. The X-Y plot between the pair of these emissions measured simultaneously to each other is thus a straight line. In addition, the extrapolated value of the X-Y plot represents a relative value for the direct excitation of PS II (F693) and/or PS I (F-718). Measurements of F-693 and F-718 induction at 70 K can thus give information not only on a) coupling of energy transfer between the two photosystems, but also b) on the magnitude of direct excitation of either one of them. In *Chlamydomonas reinhardii* (Fig.2) growth on acetate slightly decreases PS II/PS I-coupling, excitation

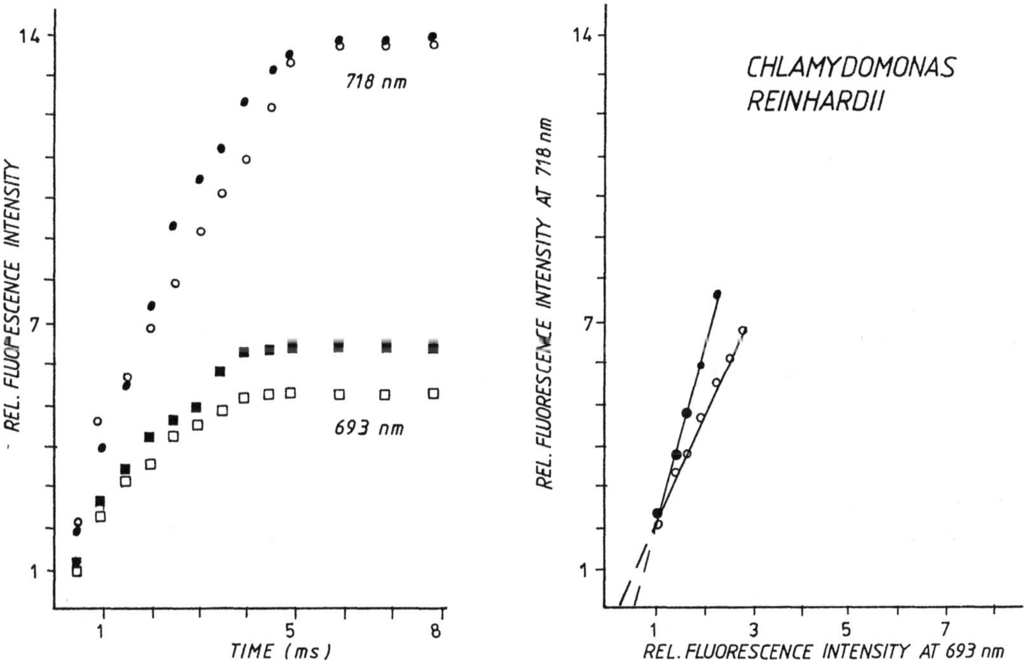

Fig.2 : Fluorescence induction measured simultaneously at 693 and 718 nm from samples of *Chlamydomonas reinhardii* frozen to -196° C after cultivation under autotrophic (closed symbols) respectively photoheterotrophic (open symbols) growth conditions.
X-Y plot of fluorescence at 693 and 718 nm

of PS I is very low. In contrast, a remarkably high excitation is seen for *Chlamydobotrys stellata* (Fig. 3). The extrapolation of the F-693/F-718 plot cuts the F-718 axis, yet there is still a spillover. Photoheterotrophic growth conditions not only increase the direct excitation of PS I but also diminish the possibility for a spillover.

The F-693/F-718 plot does not give a straight line. These results together with those presented above can best be explained by assuming a PS I connected light-harvesting system to be present in the acetate cultivated *Chlamydobotrys stellata* which provides exclusively PS I with excitation energy.

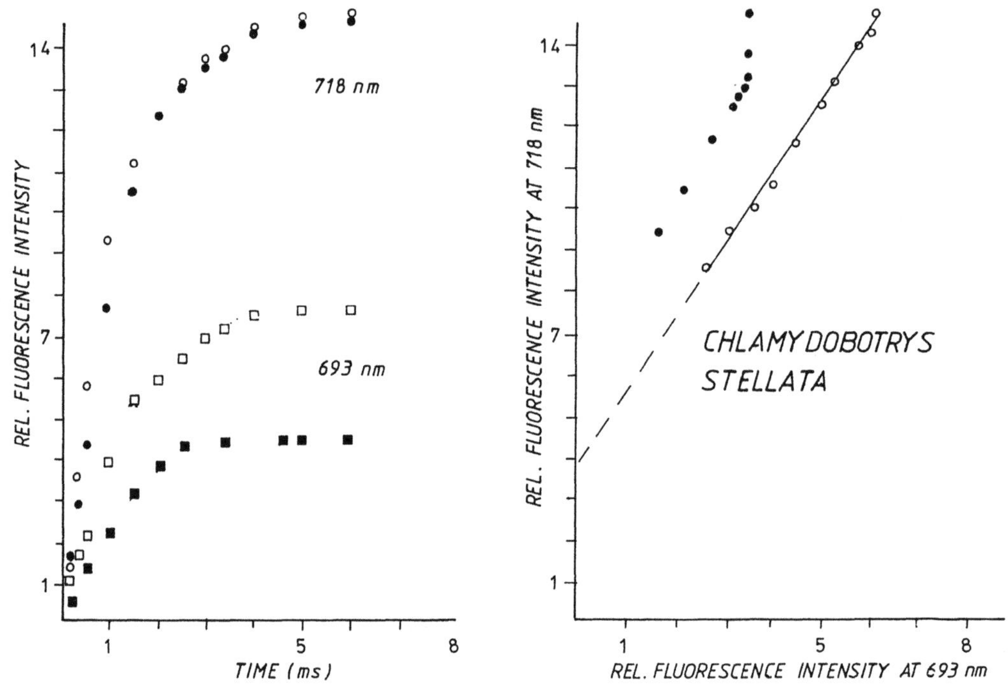

Fig.3 : Fluorescence induction at 693 and 718 nm measured simultaneously from samples of *Chlamydobotrys stellata* frozen to -196°C after cultivation under autotrophic (open symbols) respectively photoheterotrophic (closed symbols) growth conditions.
X-Y plots of fluorescence at 693 and 718 nm

Photoheterotrophic *Chlamydobotrys stellata* is characterized by its low chlorophyll b content and a remarkably high absorption in the red part of the absorption spectrum above 680 nm (Wiessner and Amelunxen 1969), Wiessner and French 1970). Aktive PS I particles can be isolated from its thylakoids which are composed out of about 11 polypeptides (Fig. 4; Brandt et al. 1983). Two of them are associated with chlorophyll a only. Their apparent molecular weights are at about 140

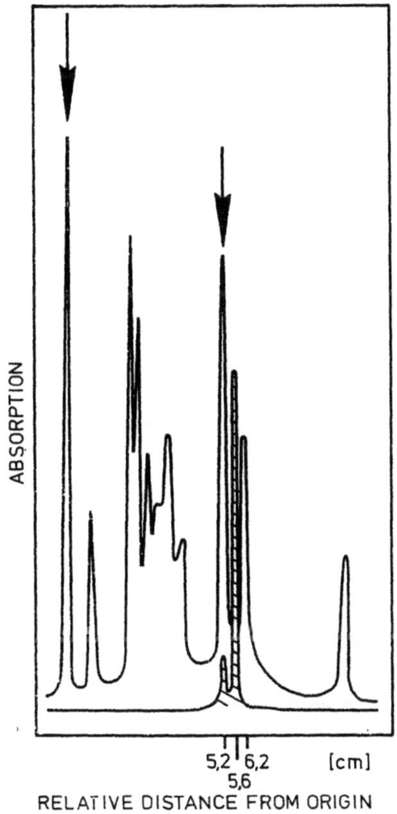

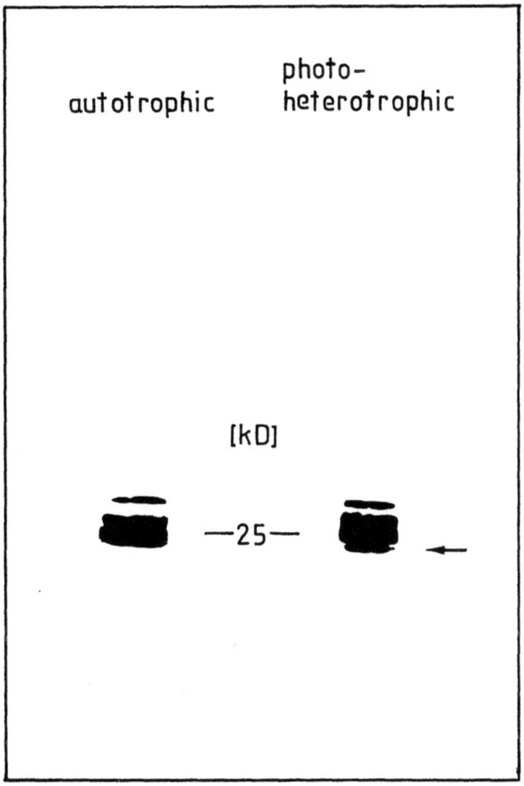

Fig.4 : Polypeptide pattern of PS I particles isolated from *Chl. stellata* by the procedure of Mullet et al.(1980). Arrows indicate chlorophyll a containing polypeptides. Dashed band = chlorophyll protein complex after heat denaturation.

Fig. 5 : Immunoblotting of thylakoid proteins from autotrophically respectively from photoheterotrophically grown *Chl. stellata* with antiserum raised to LHC.

kD and 32 kD. Denaturation of the latter by heat treatment results in a single 24 kD polypeptide. The 140 kD polypeptide represents the native PS I. The molecular weight of the 32 kD polypeptide is within the range of the normal LHC-polypeptides. It is therefore a candidate for a light-harvesting chlorophyll a-polypeptide specifically associated with PS I (LHCa). This was proven by immunoblotting technique (Fig. 5). Three LHC-polypeptides can be determined among the polypeptides of photoautotrophic cells, whereas there are four present in thylakoids of photoheterotrophically grown algae. The additional

polypeptide has a molecular weight of about 24 kD which is in good agreement to the low molecular weight chlorophyll a polypeptide of the PS I particles. It follows that the LHCa contains only one polypeptide in contrast to at least three polypeptides of LHC II.
Further prove for the specificity of LHCa comes from peptide mapping after proteolytic degradation of LHC isolated from photoheterotrophically grown algae (Fig. 6). Even though there is a great deal of conformity between the two sets of fragments. They differ from each other by the presence of three additional fragments in the case of the photoheterotrophic algae.

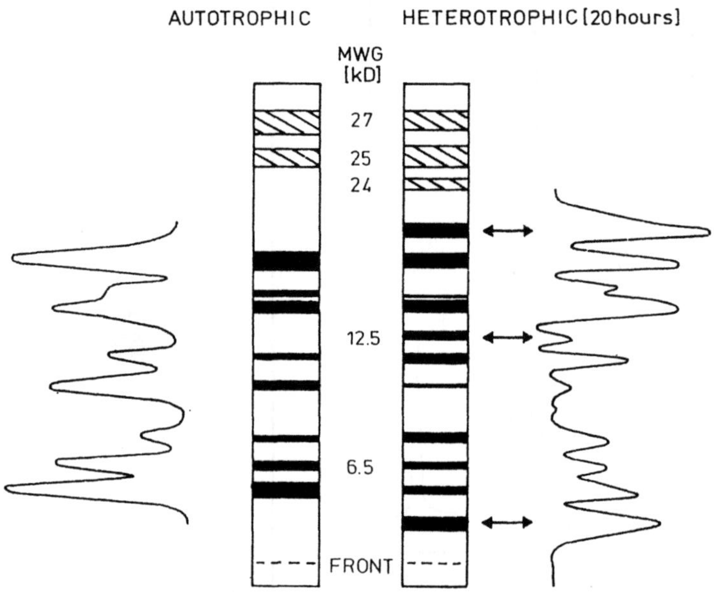

<u>Fig. 6</u> : Effects of *Staphylococcus* protease digestion on LHC isolated from autotrophically respectively from photoheterotrophically grown *Chlamydobotrys stellata*. SDS-PAGE and densidometric tracing. Arrows indicate fragments present only in the photoheterotrophic alga

To obtain information about the regulation of the synthesis of LHCa and LHC II in *Chlamydobotrys stellata*, poly(A^+)-containing m-RNA was extracted from the cells and translated in a wheat-germ system. Out of the large number of synthesized polypeptides, in the case of autotrophic organisms three immunoprecipitate with antibodies against LHC II (Fig.7) corresponding to the number of LHC II-polypeptides found in their thylakoids in vivo. Immunoprecipitation, however, yields four LHC-polypeptides when the poly (A^+) m-RNA is extractad from *Chlamydobotrys stellata* cultivated on acetate. The additional polypeptide has an apparent molecular weight of 30 kD. Antibody raised against LHCa precipitates LHCa and the 32 kD main band of LHC II. Similar to LHC II, the polypeptide of LHCa is also coded for in the nucleus and synthesized in the cytosol as a soluble precursor.

Regulation and kinetics of the LHC m-RNAs of *Chlamydobotrys stellata* were tested by isolation and translation of poly (A^+)-containing m-RNA after various periods of photoheterotrophic growth (Fig.8). The level of translatable m-RNAs for LHC II remains unchanged all over the experimental period of 20 h. The LHCa precursor not being present among the translation products of poly (A^+) m-RNA from autotrophic algae, is detectable in small quantities after 2 h of photoheterotrophic cultivation of the algae. After longer periods of photoheterotrophy the level of translatable m-RNA for the LHCa-apoprotein-precursor increases rapidly with a maximum of gene expression at the 6th h after the replacement of carbon dioxide by acetate. The amount of translatable LHCa m-RNA is reduced later on and remains constant during photoheterotrophic growth periods longer than 20 h. The constancy in the cellular levels of LHC II coding m-RNAs does not correspond to the actual amounts of the LHC II apoproteins in vivo. Their de novo synthesis decreases rapidly under photoheterotrophic growth conditions (Brandt and Wiessner 1984). It follows that one of the regulating processes of the synthesis of LHC II apoproteins is on the translational level. As the amount of m-RNA coding for the apoprotein of LHCa is comparable with the cellular content of this pro-

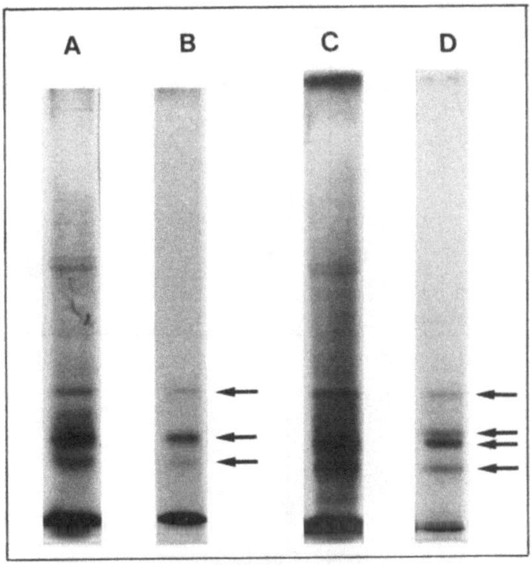

Fig. 7 : Identification of the precursors to polypeptides of LHC complexes. The translation products of poly (A^+) m-RNAs were separated by gel electrophoresis. A = Total translation products from autotrophic, C = from photoheterotrophic *Chlamydobotrys stellata*. B = Immunoprecipitation of (A), D = immunoprecipitation of (C) using antibodies raised to LHC II

tein in vivo, the regulation of LHCa synthesis apparently takes place at the transcriptional and/or the posttranscriptional level.

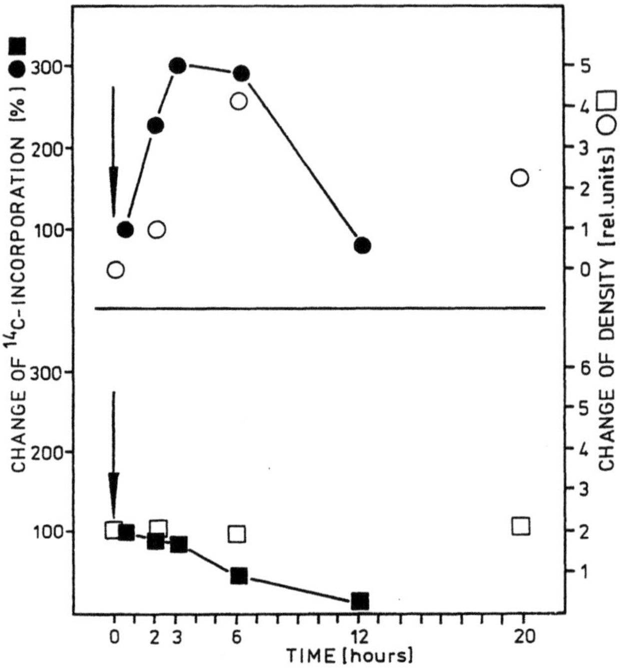

Fig.8 : Comparison of in vivo ^{14}C-incorporation into LHC II (closed squares) and LHCa (closed circles) of *Chlamydobotrys stellata* after nutritional change from autotrophy to photoheterotrophy with the relative amounts of in vitro translation of LHC II (open squares) and LHCa (open circles) precursors from poly (A^+) m-RNA

Acknowledgement

This work was supported by the Deutsche Forschungsgemeinschaft, grant Wi 191/18.

References

Allen, J.F., Bennett, J., Steinback, K.E., and Arntzen, C.J. (1981) Nature (London) 291, 25-29

Allen, J.,F., and Horton, P. (1981) Biochim. Biophys. Acta 638, 290-295

Apel, K., and Kloppstech, K. (1978) Eur. J. Biochem. 85, 581-588

Bennett, J. (1979) FEBS Lett. 103, 342-344
Bennett, J. (1983) Biochem. J. 212, 1-13
Bennett, J., Steinback, K.E., and Arntzen, C.J. (1980) Procl. Natl. Acad. Sci. U.S.A. 77, 5253-5257
Bonner, W.M., and Laskey, R.A. (1974) Eur. J. Biochem. 46, 83-88
Butler, W.L., and Strasser, R.J. (1978) In : Proc. IV. Intern. Congr. Photosynthesis 1977, eds. Hall, D.O., Combs, J., Goodwin, T.W., The Biochem. Soc., London, pp. 11-21
Brandt, P. (1980) Z. Pflanzenphysiol. 100, 95-105
Brandt, P., and Wiessner, W. (1984) Plant Physiol. 75, 253-254
Brandt, P., Zufall, E., and Wiessner, W. (1983) Plant Physiol. 71, 128-131
Cleveland, D.W., Fischer, S.G., Kirschner, M.W,, and Laemmli, U.K. (1977) J. Biol. Chem. 252, 1102-1106
Gaffron, H., Wiessner, W., and Homann, P. (1963) In : Photosynthesis Mechanisms in Green Plants. Publ. 1145, Natl. Acad. Sci.-Natl. Res. Counc., pp. 436-440
Harnischfeger, G., and Codd, G.A. (1977) Br. Phycol. J. 12, 225-232
Harnischfeger, G., and Codd, G.A. (1978) Biochim. Biophys. Acta 502, 507-513
Horton, P., Allen, J.F., Black, M.T., and Bennett, J. (1981) FEBS Lett. 125, 193-196
Kessler, S.W. (1976) J. Immunol. 117, 1482-1490
Kuhl, A. (1962) In : Beiträge zur Physiologie und Morphologie der Algen. Vorträge aus dem Gesamtgebiet der Botanik, N.F., Fischer, Stuttgart, pp. 157-166
Laemmli, U.K. (1970) Nature (London) 227, 680-685
Mans, R.J., and Norelli, G.D. (1961) Arch. Biochem. Biophys. 94, 48-53
Markwell, J.P., Thornber, J.P., and Boggs, R.T. (1979) Proc. Natl. Acad. Sci. U.S.A. 76, 1233-1235
Mende, D., and Wiessner, W. (1985) J. Plant Physiol. 118, 259-266
Mullett, J.E., Burke, J. J., and Arntzen, C.J. (1980) Plant Physiol. 65, 814-822
Neville, D.M. (1971) J. Biol. Chem. 246, 6328-6334
Pringsheim, E.G., and Wiessner, W. (1960) Nature (London) 188, 919-921
Radunz, A. (1975) Z. Naturforsch. 30c, 484-488
Roberts, B.E., and Paterson, B.M. (1973) Proc. Natl. Acad. Sci.U.S.A. 70, 2330-2334
Steinback, K.E., Bose, S., and Kyle, D.J. (1982) Arch. Biochem. Biophys. 216, 356-361
Towbin, H., Staehlin, T., and Jordan, J. (1979) Proc. Natl. Acad. Sci. U.S.A. 76, 4350-4354
Wiessner, W. (1965) Nature (London) 205, 56-57
Wiessner, W. (1968) Planta (Berl.) 79, 92-98
Wiessner, W., and Amelunxen, F. (1969) Arch. Mikrobiol. 66, 14-24
Wiessner, W., and French, C.S. (1970) Planta (Berl.) 94, 78-90

T A X O N O M I C   I N D E X

SUBJECT INDEX

MIX
Papier aus verantwortungsvollen Quellen
Paper from responsible sources
FSC® C105338

If you have any concerns about our products,
you can contact us on
ProductSafety@springernature.com

In case Publisher is established outside the EU,
the EU authorized representative is:
Springer Nature Customer Service Center GmbH
Europaplatz 3, 69115 Heidelberg, Germany

Printed by Libri Plureos GmbH
in Hamburg, Germany